T0176652

MODELING HUMAN–SYSTEM INTERACTION

MODELING HUMAN–SYSTEM INTERACTION

Philosophical and Methodological Considerations, with Examples

THOMAS B. SHERIDAN

Copyright © 2017 by John Wiley & Sons, Inc. All rights reserved

Published by John Wiley & Sons, Inc., Hoboken, New Jersey
Published simultaneously in Canada

No part of this publication may be reproduced, stored in a retrieval system, or transmitted in any form or by any means, electronic, mechanical, photocopying, recording, scanning, or otherwise, except as permitted under Section 107 or 108 of the 1976 United States Copyright Act, without either the prior written permission of the Publisher, or authorization through payment of the appropriate per-copy fee to the Copyright Clearance Center, Inc., 222 Rosewood Drive, Danvers, MA 01923, (978) 750-8400, fax (978) 750-4470, or on the web at www.copyright.com. Requests to the Publisher for permission should be addressed to the Permissions Department, John Wiley & Sons, Inc., 111 River Street, Hoboken, NJ 07030, (201) 748-6011, fax (201) 748-6008, or online at http://www.wiley.com/go/permissions.

Limit of Liability/Disclaimer of Warranty: While the publisher and author have used their best efforts in preparing this book, they make no representations or warranties with respect to the accuracy or completeness of the contents of this book and specifically disclaim any implied warranties of merchantability or fitness for a particular purpose. No warranty may be created or extended by sales representatives or written sales materials. The advice and strategies contained herein may not be suitable for your situation. You should consult with a professional where appropriate. Neither the publisher nor author shall be liable for any loss of profit or any other commercial damages, including but not limited to special, incidental, consequential, or other damages.

For general information on our other products and services or for technical support, please contact our Customer Care Department within the United States at (800) 762-2974, outside the United States at (317) 572-3993 or fax (317) 572-4002.

Wiley also publishes its books in a variety of electronic formats. Some content that appears in print may not be available in electronic formats. For more information about Wiley products, visit our web site at www.wiley.com.

Library of Congress Cataloging-in-Publication Data:

Names: Sheridan, Thomas B., author.
Title: Modeling human-system interaction : philosophical and
    methodological considerations, with examples / Thomas B. Sheridan.
Description: Hoboken, New Jersey : John Wiley & Sons, [2017] |
    Series: Stevens Iinstitute series on complex systems and enterprises |
    Includes bibliographical references and index.
Identifiers: LCCN 2016038455 (print) | LCCN 2016051718 (ebook) |
    ISBN 9781119275268 (cloth) | ISBN 9781119275299 (pdf) |
    ISBN 9781119275282 (epub)
Subjects: LCSH: Human-computer interaction. |
    User-centered system design.
Classification: LCC QA76.9.H85 S515 2017 (print) |
    LCC QA76.9.H85 (ebook) | DDC 004.01/9–dc23
LC record available at https://lccn.loc.gov/2016038455

Cover Image: Andrey Prokhorov/Gettyimages

Set in 11/13pt Times by SPi Global, Pondicherry, India

Printed in the United States of America

10 9 8 7 6 5 4 3 2 1

CONTENTS

PREFACE

This book has evolved from a professional lifetime of thinking about models and, more generally, thinking about thinking. I have previously written seven books over a span of 42 years, and they all have all talked about models, except for one privately published as a memoir for my family. One even dealt with the concept of God and whether God is amenable to modeling (mostly no). So what is new or different in the present book?

The book includes quite a bit of the philosophy of science and the scientific method as a precursor to discussing human–system models. Many aspects of modeling are discussed: the purpose and uses of models for doing science and thinking about the world and examples of different kinds of models in what has come to be called *human–system interaction* or *cognitive engineering*. Along with new material, the book also includes many modeling ideas previously discussed by the author. When not otherwise cited, illustrations were drawn by the author for the book or were original works under the author's copyright or previously declared by the author to be in public domain prior to publication.

I gratefully acknowledge contributions to these ideas from many colleagues I have worked with, especially Neville Moray, who has been my friend and invaluable critic over the years, and Bill Rouse, who shepherded the book as Wiley series editor. Modeling contributions of past coauthors Russ Ferrell, Bill Verplank, Gunnar Johannsen, Toshi Inagaki, Raja Parasuraman, Chris

Wickens, Peter Hancock, Joachim Meyer, and many other colleagues and former graduate students are gratefully acknowledged.

Finally, I dedicate this effort to Rachel Sheridan, my inspiration and life partner for 63 years.

INTRODUCTION

This is a book about models, scientific models, of the interaction of individual people with technical environments, which has come to be called *human–system interaction* or *cognitive engineering*. The latter term emphasizes the role of the human intelligence in perceiving, analyzing, deciding, and acting rather than the biomechanical or energetic interactions with the physical environment

Alphonse Chapanis (1917–2002) is widely considered to be one of the founders of the field of human factors, cognitive engineering, or whatever term one wishes to use. He coauthored one of the (if not THE) first textbooks in the field (Chapanis et al., 1949). I had the pleasure of working with him on the original National Research Council Committee in our field (nowadays called Board on Human Systems Integration, originally chaired by Richard Pew). I recall that Chapanis, while a psychologist by training, repeatedly emphasized the point that our field is ultimately applied to *designing technology* to serve human needs; in other words it is about engineering. Models are inherent to doing engineering.

More generally, models are the summaries of ideas we hang on to in order to think, communicate to others, and refine in order to make progress in the world. They are *cognitive handles*. Models come in two varieties: (1) those

Modeling Human–System Interaction: Philosophical and Methodological Considerations, with Examples, First Edition. Thomas B. Sheridan.
© 2017 John Wiley & Sons, Inc. Published 2017 by John Wiley & Sons, Inc.

couched in language we call *connotative* (metaphor, myth other linguistic forms intended to motivate a person to make his or her own interpretation of meaning based on life experience) and (2) language we call *denotative* (where forms of language are explicitly selected to minimize the variability of meaning across peoples and cultures). Concise and explicit verbal statements, graphs, and mathematics are examples of denotative language. There is no doubt that connotative language plays a huge role in life, but science depends on denotative expression and models couched in denotative language, so that we can agree on what we're talking about.

The book focuses on the interaction between humans and systems in the human environment of physical things and other people. The models that are discussed are representations of events that are observable and measurable. In experiments, these necessarily include the causative factors (inputs, independent variables), the properties of the human operator (experimental subject), the assigned task, and the task environment. They also include the effects (outputs, dependent variables), the measures of human response correlated to the inputs.

Chapters 1–3 of the book are philosophical, and apply to science and scientific models quite generally, models in human–system interaction being no exception. Chapter 1 begins with a discussion of what knowledge is and what the scientific method is including the philosophical distinction between private (subjective) knowledge and public (objective) knowledge, the importance of doubt, using and avoiding evidence, objectivity and advocacy, bias, analogy, and metaphor.

Chapter 2 defines the meaning of "model" and offers a six-factor taxonomy of model attributes. It poses the question of what is to be gained by modeling and the issue of social (group) behavior (choice) as related to that of the individual.

Chapter 3 discusses various distinctions in modeling: those between objective and subjective data, simple and complex models, descriptive and prescriptive models, static and dynamic models, deterministic and probabilistic models, level of abstraction relative to the target thing, or events being characterized, and so on.

Chapter 4 describes various forms of model representation and provides examples: modeling in words, graphs, maps, diagrams, logic diagrams, symbols (mathematic equations), and statistics.

Chapters 5–8 offer specific examples of cognitive engineering models applicable to humans interacting with systems. In each case, a brief discussion is provided to explain the gist of the model. These chapters are in the sequence popularly ascribed to what are known as the four sequential *stages* of how humans do tasks (and the chapters are so titled): acquiring information, analyzing the information, deciding on action, and implementing and evaluating the action. There is no intention to be comprehensive in the selection of these models. Many of the models discussed are those that I have had a hand in developing, so that admittedly there is a bias. I have tried to include other authors' models so as to provide a variety and to populate the four model categories representing the accepted taxonomy of four sequential stages for a human operator doing a given task.

Chapter 9 deals with the many aspects of human–automation interaction, wherein some of the functions of information acquisition, analysis, decision-making, and action implementation are performed by machine, and the human is likely to be operating as a supervisor of the machine (computer) rather than being the only intelligence in the system.

Chapter 10 takes up the issue of mental modeling: representing what a person is thinking, which is not directly observable and measurable, and so must be inferred from what a subject says or does in a contrived experiment. A mental model is surely not a scientific model in the same sense as those covered in previous chapters, yet the cognitive scientists working on mental modeling would surely claim that they are doing science by inferential measurement of what is in the head. The chapter provides four very different and contrasting types of mental model.

Chapter 11 deals with modeling of large-scale societal issues including health care, climate change, transportation, jobs and automation, wealth inequity, privacy and security, population growth, and governance. Such issues and the associated macromodels involve a large number of people. The book cannot possibly cope with reviewing the many such models that exist. Instead, the purpose of the chapter is to ask whether and how our cognitive engineering micro-models relate to the societal macro-models. Is there a connection that can be useful? Do or can the human performance models scale up to predict how larger groups of people, for example, at the family, tribe, region, nation, or world, interact with their technology and physical environments? Is our modeling community remiss in not contributing more to these larger-scale modeling efforts? Are there some specific human factors or cognitive engineering aspects that we can help with?

Finally the Appendix gives the mathematical particulars of selected models for which there is a sound mathematic basis. Thus the reader of the main text can skip the mathematical details or refer to them as convenient. However, I add the caveat that insofar as quantification is appropriate and warranted based on relevant empirical data, such quantitative models are predictive and thus more useful for engineering purposes.

I must reiterate that the models I have selected are necessarily ones that I am familiar with. I make no pretense that the selection is even handed, and it certainly is not comprehensive. The reader may feel that some models are dated and no longer in fashion (yes models can have runs of fashion), though I would maintain that all those included have passed the test of time and continue to have relevance.

As noted before, the book focuses on what has come to be called human–system interaction or cognitive engineering models. It does not include biomechanical models and kinematic or energetic aspects of humans performing tasks, which subfield is commonly associated with "ergonomics." This is surely a shortcoming for a book purporting to be about "human–system interaction." I would simply assert that for most of the tasks implied by the example models, the constraints of biomechanics are not major factors in task performance. Further, I have barely mentioned artificial intelligence, robotics, computer science, control theory, or other engineering theories that are emerging as companion fields to human–automation interaction.

1

KNOWLEDGE

GAINING NEW KNOWLEDGE

Knowledge can be acquired by humans in many ways. Surely, there are also many ways to classify the means to acquire knowledge. Here are just a few ways.

One's brain can acquire knowledge during the evolutionary process by successive modifications to the genes. That finally results in fertilization of egg by sperm and the gestation process in the mother. Certainly, all this depends on the sensory–motor "operating system" software that makes the sense organs and muscles work together. But evolution also plays at the level of higher cognitive function. As Noam Chomsky has shown us (Chomsky, 1957), much of the syntactic structure of grammar is evidently built in at birth. What knowledge we acquire after birth is a function of what we attend to, and what we attend to is a function of our motivation for allocating our attention, which ultimately is a function of what we know, so knowledge acquisition after birth is a causal circle.

Learning has to do with how we respond to the stimuli we observe. Perhaps, the oldest theory of learning is the process of Pavlovian (classical) conditioning, where a stimulus, originally neutral in its effect, becomes a

Modeling Human–System Interaction: Philosophical and Methodological Considerations, with Examples, First Edition. Thomas B. Sheridan.
© 2017 John Wiley & Sons, Inc. Published 2017 by John Wiley & Sons, Inc.

signal that an inherently significant (reward or punishment) unconditioned stimulus is about to occur. This results only after multiple pairings, and the brain somehow remembers the association. The originally neutral stimulus becomes conditioned, meaning that the person (or animal) responds reflexively to the conditioned stimulus the same as the person would respond to the unconditioned stimulus (e.g., the dog salivates with the light or bell).

A different kind of learning is Skinnerian or *operant* conditioning (Skinner, 1938). This is where a voluntary random action (called a free operant) is rewarded (reinforced), that association is remembered, and after sufficient repetitions, the voluntary actions occur more often (if previously rewarded). Operant learning can be maintained even when rewards are infrequently paired with the conditioned action.

There are many classifications of learning (http://www.washington.edu/doit/types-learning). Bloom et al. (1956) developed a classification scheme for types of learning which includes three overlapping domains: cognitive, psychomotor, and affective. Skills in the cognitive domain include *knowledge* (remembering information), *comprehension* (explaining the meaning of information), *application* (using abstractions in concrete situations), *analysis* (breaking down a whole into component parts), and *synthesis* (putting parts together to form a new and integrated whole).

Gardner (2011) developed a theory of multiple intelligences based upon research in the biological sciences, logistical analysis, and psychology. He breaks down knowledge into seven types: *logical–mathematical intelligence* (the ability to detect patterns, think logically, reason and analyze, and compute mathematical equations), *linguistic intelligence* (the mastery of oral and written language in self-expression and memory), s*patial intelligence* (the ability to recognize and manipulate patterns in spatial relationships), *musical intelligence* (the ability to recognize and compose music), *kinesthetic intelligence* (the ability to use the body or parts of the body to create products or solve problems), *interpersonal intelligence* (the ability to recognize another's intentions and feelings), and *intrapersonal intelligence* (the ability to understand oneself and use the information to self-manage).

Knowledge can be public, where two or more people agree on some perception or interpretation and others can access the same information. Or it can be private, where it has not or cannot be shared. The issue is tricky, and that is why modelability is proposed as a criterion for what can be called public knowledge. Two people can look at what we call a red rose, and agree that it is red, because they have learned to respond with the word *red* upon observing that stimulus. But ultimately exactly what they experienced cannot be shared, hence is not public.

We can posit that some learning is simply accepting, unquestioningly, information from some source because that source is trusted or because the learner is compelled in some way to learn. We finally contrast the aforementioned models to learning by means of the scientific method, which is detailed in the following text. Critical observation and hypothesizing are followed by collection of evidence, analysis, logical conclusions, and modeling to serve one's own use or to communicate to others.

SCIENTIFIC METHOD: WHAT IS IT?

How to determine the truth? Science has its own formal model for this. The scientific method is usually stated as consisting of nine steps as follows:

1. Gather information and resources (informal observation).
2. Question the relationships between aspects of some objects or events, based on observation and contemplation. An incipient mental model may already form in the observer's head.
3. Hypothesize a conjecture resulting from the act of questioning. This can be either a predictive or an explanatory hypothesis. In either case, it should be stated explicitly in terms of independent and dependent variables (causes and effects).
4. Predict the logical consequences of the hypothesis. (A model will begin to take shape.)
5. Test the hypothesis by doing formal data collection and experiments to determine whether the world behaves according to the prediction. This includes taking pains to design the data-taking and the experiment to minimize risks of experimental error. It is critical that the tests be recorded in enough detail so as to be observable and repeatable by others. The experimental design will have a large effect on what model might emerge.
6. Analyze the results of the experiment and draw tentative conclusions. This often involves a secondary hypothesis step, namely exercising the statistical *null hypothesis*. The null hypothesis is that some conjecture about a population of related objects or events is false, namely that observed differences have occurred by chance, for example, that some disease is not affected by some drug. Normally, the effort is to show a degree of statistical confidence in the failure and thus rejection of the null hypothesis. In other words, if there is enough confidence that the differences did not occur by chance, then the conjectured relationship exists.

7. Draw formal conclusions and model as appropriate.

8. Communicate the results, conclusions, and model to colleagues in publication or verbal presentation, rendering the model in a form that best summarizes and communicates the determined relationships.

9. Retest and refine the model (frequently done based on review and critique by other scientists).

FURTHER OBSERVATIONS ON THE SCIENTIFIC METHOD

The scientific method described earlier is also called the *hypothetico-deductive* method. As stated, it is an idealization of the way science really works, as the given scientific steps are seldom cleanly separated and the process is typically messy. Often experimentation is done in order to make observations that provoke additional observations, questions, hypotheses, predictions, and rejections or refinements of the starting hypothesis. Especially at the early observation stage, the process can be very informal. One of the writer's students used to say that what we did in the lab was "piddling with a purpose." Einstein is said to have remarked that the most important tool of the scientist is the wastebasket.

Philosopher statesman Francis Bacon (1620) asserted that observations must be collected "without prejudice." But as scientists are real people, there is no way they can operate free of some prejudice. They start with some bias as to their initial knowledge and interests, their social status and physical location, and their available tools of observation. They are initially prejudiced as to what is of interest, what observations are made, and what questions are asked.

Philosopher Karl Popper (1997) believed that all science begins with a prejudiced hypothesis. He further asserted that actually a theory can never be proven correct by observation, but it can only be proven incorrect by disagreement with observation. Scientific method is about falsifiability. That is the basis of the null hypothesis test in statistics. (But, of course, the falsifiability is itself subject to statistical error; one can only reject the null hypothesis with some small chance of being wrong.) The American Association for the Advancement of Science asserted in a legal brief to the U.S. Supreme Court (1993) that "Science is not an encyclopedic body of knowledge about the universe. Instead, it represents a process for proposing and refining theoretical explanations about the world that are subject to further testing and refinement."

Historian Thomas Kuhn (1962) offered a different perspective on how science works, namely, in terms of *paradigm shifts*. Whether in psychology or cosmology, researchers seem to make small and gradual refinements of accepted models, until new evidence and an accompanying model provokes a radical shift in paradigm, to which scientists then adhere for a time. When a new paradigm is in process of emerging the competition between models and their proponents can be fierce, even personal (who discovered X first, who published first, whose model offers the best explanation). We also must admit that search for truth is not the only thing that motivates us as scientists and modelers. We are driven by ambition for recognition from our peers as well as by money.

The idea of *reproducible observability* deserves emphasis. Having to deal with observables is the most critical factor in an epistemological sense (what we know). This is because it distinguishes what may be called truth based on scientific evidence that is openly observable from experiences that are not observable by others (e.g., personal testimony and anecdotal evidence). Observability also comes into play for what are called mental models.

Mental models can be called models of a sort, but being private they are not subject to direct observation by other people. Experiments in psychophysics, where subjects make verbal category judgments or button-push responses to physical stimuli of sound, light, and so on, are regarded as conforming to scientific method. This is because the human is making a direct mechanical response to a given stimulus, such as pushing a button, not having to articulate in arbitrary words what the human is thinking. However, when subjects are asked to explicate in their own words their mental models of how they believe something works, or what are cognitive steps of a particular task as might be asked of subject matter experts, there is no external physical reference; scientific method here is more challenging. Of course, there must be repeatability or aggregation of the results from many subjects. Observability clearly is a challenge for modeling.

While social scientists often point out that humans have a predilection for reaffirming the *status quo*, science nevertheless is a truth system committed to change, as warranted, rather than preservation. But while science actively pursues possibilities of change, the null hypothesis testing by its very nature demands a significant level of statistical confidence in order to reject the null hypothesis that there is no real change (that there is only random difference between the hypothesized variant and the control).

The scientific method is a method wherein inquiry regards itself as fallible and purposely probes, criticizes, corrects, and improves itself. This universally accepted attribute stands in sharp contrast to religious and political traditions around the world. Science is the one human endeavor that has proven relatively immune to the passions that otherwise divide us.

REASONING LOGICALLY

One can derive a model by logical reasoning or it may be stated out of ignorance or for purposes of deception. It may also be a metaphorical model, where, because of ambiguity in the words or drawings, it is not possible to conclude that it is logical. If it is based on logic and assuming no ambiguity in the words, pictures, or symbols, it is relevant to mention three different types of logical reasoning: deduction, induction, and abduction.

Deduction allows deriving b from a only where b is a formal logical consequence of a. In other words, deduction is the process of deriving the consequences of what is assumed. Given the truth of the assumptions, a valid deduction guarantees the truth of the conclusion. For example, given that all bachelors are unmarried males, and given that some person is a bachelor, it can be deduced that that person is an unmarried male.

Induction allows inferring b from a, where b does not follow necessarily from a. The existence of a might give us very good reason to accept b, but it does not ensure that b is true. For example, if all of the swans that we have observed so far are white, we may induce that all swans are white. We have good reason to believe the conclusion from the premise, but the truth of the conclusion is not guaranteed. (Indeed, it turns out that some swans are black.)

Abduction allows inferring a as an explanation of b. Put another way, abduction allows the precondition a to be abduced from the consequence b. Deduction and abduction thus differ in the direction in which a rule like "a entails b" is used for inference. As such, abduction is formally equivalent to a logical fallacy that a unique a occurs where actually there are multiple possible explanations for b. For example, after glancing up and seeing the eight ball moving in some direction we may abduce that it was struck by the cue ball. The cue ball's strike would account for the eight ball's movement. It serves as a hypothesis that explains our observation. There are in fact many possible explanations for the eight ball's movement, and so our abduction does not leave us certain that the cue ball did in fact strike the eight ball, but our abduction is still useful and can serve to orient us in our surroundings. This process of abduction is an instance of the scientific method. Logically, there are infinite possible explanations for any of the physical processes we observe, but from our experience we are inclined to abduce a single explanation (or a few explanations) for them in the hopes that we can better orient ourselves in our surroundings and then eliminate some of the possibilities by further observation or experiment.

PUBLIC (OBJECTIVE) AND PRIVATE (SUBJECTIVE) KNOWLEDGE

Model development necessarily begins in the head of some person, and only later is the model rendered in words, graphics, or mathematical language so that it can be communicated to others. The term *mental model* refers broadly to a person's private thoughts, though some psychologists would confine its use to well-formed ideas about the structure or function of some objects or events, such that it is potentially communicable in understandable format to another person.

Psychologists have struggled for many years with how to extract a person's mental model. Psychophysical methods of having experimental subjects' rank-order stimuli or assign preferential numbers or descriptive categories (as done by political pollsters) are a common approach. With more complex situations, and particularly those that have social implications, a problem is that what people say they believe and what they actually believe may be quite different. People are inclined to say what they think some other person wants to hear; social etiquette reinforces this behavior. Later in the book, Chapter 10 deals with mental models in more detail.

THE ROLE OF DOUBT IN DOING SCIENCE

For many years, philosophers, from Rene Descartes to Charles Sanders Peirce (2001), have proposed *methodological doubting* (also called *Cartesian skepticism*, *methodological skepticism*, or *hyperbolic doubt*) as a means to test the truth of one's beliefs. I recall attending a "Skeptics Seminar" at MIT at which the famous mathematician and founder of cybernetics Norbert Wiener emphasized the importance of doubting in order to refine one's beliefs.

Doubt is deliberation on error and failure. Together with a graduate student, I once offered an experimental graduate course called "Seminar on Failure" in which historians, psychiatrists, scientists, and engineers were invited to reminisce on failures in their own professional experiences. We learned that failure often led to discovery and learning, and that success often led to overconfidence and carelessness and eventually to failure. Failure and success are the Taoist yin and yang of our experience; they are mutually complementary; one could not exist without the other.

Doubt and questioning motivate scientists to think of different hypotheses and try different experiments. Pioneer in cybernetics Ross Ashby (1956) is known for his "law of requisite variety," which asserts that in order for one system to control another it must possess a greater variety of states. This law has been applied to genetic mutations believed to be essential to the evolution of species as per Darwin's theory. There is requisite variety in our own social and environmental encounters in life, from which we can refine our beliefs and models of what works and what does not work. Edwin Hubble's doubt led to the discovery that the brightness of a spot in the Andromeda cluster could not be from our own galaxy, the first proof that there were other galaxies in the universe. Sometimes, it is not simply doubt but absolute failure that provokes a wider search for solution and improvement, whether in a computer or a person.

EVIDENCE: ITS USE AND AVOIDANCE

It might be assumed that rational people accept a model or a belief because the preponderance of evidence supports that model or belief. But people are always free to "choose" a belief because it is what parents or peers say they believe, or what individuals believe they should believe, for whatever reason. Perhaps, it is because they think believing will make them feel better, or there will be punishment for professing disbelief. But these are not acts of seeking truth. No one says that truth-seeking is easy, particularly when what appears to be the rational truth is in conflict with claims by authority figures or other trusted sources.

Surely, truth-seeking has its costs. In many ways, evolution has designed us not to be rational in every act. Hyper-rational insistence on bare-faced truth clearly gets us into trouble—in interpersonal relations where etiquette is required, in supporting and defending a child from destructive criticism, in winning an argument, and so on. The truth may be messy and ugly. Some will surely feel that absolute truth is second to happiness, and who will ever know the ultimate truth anyway? Isn't happiness the real goal? In any case, we may want to be careful about flaunting what we believe to be the truth.

METAPHYSICS AND ITS RELATION TO SCIENCE

Metaphysics is a traditional branch of philosophy concerned with being— what things exist and what are their properties. Prior to the eighteenth century, all questions of ultimate reality were addressed by philosophers. Aristotle believed that things have within them their own purpose (a teleology). He is well known for distinguishing between two essential properties of things: potentiality (the possibility of any property a thing can have) and actuality (what is actually in evidence). There are modern manifestations of Aristotle's dichotomy, such as the distinction between potential and kinetic energy in physics. Thomas Aquinas called metaphysics the "queen of sciences" and wrote extensively on the subject.

Ever since the Enlightenment, the field of metaphysics has evolved into a philosophical pursuit of topics that have not been easy for science to handle, such as being (existence), mind, perception, free will, consciousness, and meaning. Science has begun to confront some of these issues. Whether over time they will be clarified remains as a conundrum, or whether the verbal constructs will just fade away remains to be determined. I would guess some of each.

Descartes' famous conclusion that "I think, therefore I am" was proffered as a metaphysical basis for his own existence. He asserted that while all else could be doubted, the fact that he could do the doubting meant that he must exist in order to doubt. Such thought experiments are common in philosophy. One famous thought experiment allowed that to imagine a thing or event is to make it exist, namely the famous "proof of God" offered by Anselm of Canterbury. Surely, thoughts do exist as neural activity in the brains of people who have the thoughts; they constitute what we have called mental models. But our concern here is mostly for deducing the existence of things perceived from observables, with the exception of "mental models," which are dealt with separately in Chapter 10.

OBJECTIVITY, ADVOCACY, AND BIAS

Ideally, the scientist is supposed to be disinterested, completely impartial to how any test of a hypothesis turns out, and what the implications of the results are for the world. But scientists are people, and people do things because they are interested in achieving some objectives. The very act of formulating a hypothesis is a creative act, motivated by the interests of the scientist. But given this unavoidable level of interest, the scientist has an obligation to be as objective as possible, and not favor one aspect of the results while hiding some other aspect.

In truth, everyone is subject to bias, including this author. Psychologists have studied human biases extensively. In dealing with value-laden topics, biases are especially salient. One most pertinent category of bias is what is called the *confirmation bias*. Nickerson (1998) defines the confirmation bias as "the seeking or interpreting of evidence in ways that are partial to existing beliefs, expectations, or a hypothesis in hand." He reviews evidence of such a bias in a variety of guises and gives examples of its operation in several practical contexts. Other biases include overconfidence in one's own predictions, giving more weight to recent events as compared to earlier events in judging probabilities, ascribing greater risk to situations one is forced into as compared to those voluntarily selected, and inferring illusory causal relations.

Many democratic societies have judicial systems based on advocacy, where the advocates of the two or more sides of any argument confront one another in a highly procedural venue in front of a jury of peers. The assumption is that the jury can judge the arguments, weigh the evidence presented, and detect efforts to hide something. Because conventional jury selection is not set up to include experts on particular fields of science or technology, there have been various proposals to develop "science courts" that would somehow demand more rigorous objectivity in presenting and judging arguments and evidence.

ANALOGY AND METAPHOR

Analogy is a broad class of cognitive process or linguistic expression involving transfer of meaning from one object or event to another because of some similarity. Analogy plays a critical role in creativity, problem solving, memory, perception, explanation, and communication.

Science depends on metaphor (and simile) in the sense that an active and curious observer is constantly seeing analogies, likenesses between elements of nature. The observed likenesses lead to hypotheses that permit generalization.

For example, in the field of physics, mechanical force, pressure, electrical voltage, and temperature are all seen as having the property of effort. Mechanical velocity, fluid flow, electrical current, and heat flow are all seen as having the property of flow. Mechanical friction, fluid viscosity, electrical resistance, and thermal insulation all have the property of resistance to flow. Depending on the spatial configuration, the equations relating the forces, flows, and resistance to flow are the same (to a first approximation), namely

$$\text{Effort} = \text{flow} \times \text{resistance}$$

Such analogies are powerful concepts in physical science that also have counterparts in traffic analysis, economics, and other fields. Much academic teaching in the physical sciences and engineering employs analogy.

Metaphor is a figure of speech that makes use of analogy. It is a word or phrase describing a thing or an action that is regarded as representative or symbolic of something else, especially something abstract—even though it is not literally applicable. This is in contrast to a simile, in which something is said to be like something else, and the attribute of likeness is spelled out or implied. Joseph Campbell (1949) uses the following examples: The boy runs like a deer (simile). The boy is a deer (metaphor).

Metaphor finds its place in literature, poetry, music, and the arts; they depend on metaphor. Metaphor can say things more emotionally powerful than simple rational statements. Metaphor provokes the human imagination, as in myths, allegories, and parables. It has been said that literature and the arts often realize human truths well before other branches of human endeavor do. Since it is a figure of speech, a metaphor is not a belief (a mental event); rather it is a way to describe a belief. Things are not metaphors but can be expressed through metaphors. Every metaphor is both metaphorically true (if it is an apt description) and literally false.

An analog representation, whether in words or any other medium, can be a model. This author cut his academic teeth using analog computer operations as models of human behavior. The point is that certain relationships (e.g., in magnitude and time of the graphical traces generated by the analog computer) denote magnitude and time relationships of the target thing or events being modeled. It is understood by the user of the analog computer that other properties of the analog computer (physical hardware, electron flow) are irrelevant.

Metaphorical description can be a kind of model, though not a scientific model, as previously noted. This is because the interpretation of the metaphor as a representation of the target object or event is a creative act by the reader or listener, since the connection between the metaphor and the target is not explicit (it is connotative). For example, religious myth fits this category.

2

WHAT IS A MODEL?

In this section, we discuss the notion of modeling in very general terms. A definition is offered, a taxonomy of attributes of models is proposed by which to evaluate models, and some important distinctions in modeling are made.

What do a verbal treatise on how some aspect of the economy works, a miniature replica of an airplane, a mathematical equation characterizing driver steering behavior, and a girl posing before a camera or a painter have in common? We call each of them a model. There are other entities that we do not normally call models, such as a myth or a poem, but by a broad definition they too are models, though not the kind this book focuses on.

DEFINING "MODEL"

A model can be defined as a concise, denotative representation of the structure or function of some selected aspects of our world to one or more observers for the purposes of communicating a belief about some relationships, expressing a conjecture, making a prediction, or specifying a design of a thing or a set of events.

Modeling Human–System Interaction: Philosophical and Methodological Considerations, with Examples, First Edition. Thomas B. Sheridan.
© 2017 John Wiley & Sons, Inc. Published 2017 by John Wiley & Sons, Inc.

The term "model" as defined is quite general. Beyond the idea of representation are synonyms such as specification, rendering, map, or characterization of the relations between elements or variables of the defined set of objects or events. There are semantic overlaps with related terms such as abstraction, construction, explanation, portrayal, depiction, theory, idea, concept, paradigm, and pattern.

An important word in the given definition is *selected*. No model purports to include all of the factors (variables) relating to the target object or set of events. Much (indeed most) will be left out. If nothing were left out, then we would have a copy, not a model. So, modeling is a many-to-fewer mapping from the original to the model. It is obligatory that the modeler specifies which variables are included, and by implication everything else is omitted. The idea is to capture the independent variables that have the greatest effect on the dependent variables of interest.

A model can be said to be a description of a concept or thing. Models can be described using words, graphics (e.g., graphs, diagrams, and pictures), mathematical equations, physical things, or some combination of these (e.g., computer simulations that run equations and output numbers, graphs, and dynamic animations). The word *concise* is added to the definition to emphasize that a model states the intended relationships briefly but completely and unambiguously, eliminating redundancy and extraneous words or symbols. This excludes long-winded, wordy statements or graphics with elaboration that are unnecessary to the message. Unfortunately, there is no clean distinction as to when a particular statement qualifies as a model or not.

Semiotics is the theory of signs, which includes *syntax* (grammar, structural constraints of words or symbols), *semantics* (their meaning), and *pragmatics* (effects of their use on people). Semiotics makes an important distinction between denotation and connotation. *Denotation* refers to the explicit literal meaning of the words, symbols, or signs. *Connotation* refers to the implied or suggested meaning, as with metaphor. A photo of or a verbal statement about a red rose with a green stem denotes nothing more than a red rose with a green stem, but it connotes affection or celebration. In the book, we mostly restrict the word model to denotative representation. This accords with the use of models in all fields of science.

In modern times, we see ever-greater use of denotative models, particularly in science and technology. Dating from early Greek civilization, we have had verbal models, but more and more since the Enlightenment we have seen the development of mathematical, graphical, and other symbolic models. Our ability to run fast-time computer simulations on large databases for weather, physics, engineering, economics, genomics, transportation, and so on, has made huge strides in recent years.

A model typically states explicitly, or at least strongly implies, an "IF X, THEN Y" relationship. For example, if you look at a particular location X on a map or within the human body, you will find Y. Or if X is the input or independent variable to a given system or process, then Y will be the result (output). The latter is common to a scientific model that is intended to generalize about and predict observations.

Ideally, scientific models are intended to be predictive so as to be useful in some application. It can be said that models without application are useless (essentially by definition if interpreted broadly). It should be noted that basic science routinely develops models that are used only to communicate understanding of how the world works, and which only much later may find practical application.

Some people distinguish a model from a theory, where a model is a rendering of specific relationships relevant to a theory, a theory being a statement that circumscribes some relationships within the larger, more complex reality. The theory can be just a hypothesis or it can be an accepted statement of some reality. Ideally the theory includes statements of constraints on application: where, when, and how the theory or the derived model applies or does not apply. So, models and theories of application naturally go together, and unfortunately the terms are often used to mean the same thing.

A cross section of denotative models in different applications might be the following:

- Business model: a framework of the business logic of a firm
- Computer model: a computer program that simulates abstractions about a particular system
- Ecosystem model: a representation of components and flows through an ecosystem
- Physiological response model: description of a neural process that simulates the response of the motor system in order to estimate the outcome of a neural command
- Macroeconomic model: a representation of some aspect of a national or regional economy
- Map: used for navigation by air, sea, or land
- Mechanism model: a description of a system in terms of its constituent parts and mechanisms
- Molecular model: a physicochemical or mathematical description that describes the behavior of molecules
- Pension model: a description of a pension system including simulations and projections of assets

- Standard model of physics: the theory in particle physics that describes certain fundamental forces and particles
- Statistical model: in applied statistics, a parameterized set of probability distributions
- Wiring diagram: as used by an electrician

Chapters 5–10 provide examples of models of human–system interaction of the types prevalent in cognitive engineering.

We note that the models in Chapters 5–10 are often or usually generated from observable data of what human subjects say or do in controlled laboratory experiments, where subject responses are not expressed in free form but are highly constrained so that credible averages and statistics may be calculated by aggregating subject responses.

MODEL ATTRIBUTES: A NEW TAXONOMY

Table 2.1 illustrates a proposed general taxonomy of models (Sheridan, 2014), with six attributes shown as rows and relative levels as column headings. For each attribute, there are three-level descriptors, marked 1, 2, and 3, where level 1 is the least of the given attribute, level 3 is the most of that attribute, and

TABLE 2.1 A taxonomy of model attributes

	Attribute	1 (Least)	2 (Moderate)	3 (Most)
A	Applicability to observables	Not based on observables	Describes past observables	Predicts future observables
B	Dimensionality	Single-input– single-output	Multi-input– single-output	Multi-input– multi-output
C	Metricity	Limited to nominal relationships	Primarily ordinal relationships	Entirely cardinal relationships
D	Robustness	Unique focus on limited objects or events	Moderate focus to a variety of objects or events	Comprehensive of a wide slice of nature
E	Social penetration	Confined to a mental model	Communicated to the relevant community	Accepted and used by the relevant community
F	Conciseness	Wordy, redundant, unclear, ambiguous expression	Minor wordiness, redundancy, or ambiguity	Concise and clear, no redundancy, no ambiguity

level 2 is in between those limits. If each of the six attributes were considered to have only three discrete levels (1, 2, 3), there would be $3 \times 3 \times 3 \times 3 \times 3 \times 3 = 729$ combinations. However, for each attribute, the three levels can also be considered rough descriptors of values along a continuum, yielding what scientists would call a continuous six-dimensional state space.

(A) Applicability to Observables

Consider the first attribute, *applicability to observables*, the one most coupled to what the modeler intends the model to be. The term *observables* here refer to objects or events that can be sensed directly by humans or measured using some repeatable physical means. Also, it is important to note that the objects or events must potentially be observable by any human; personal experiences that cannot be shared with others do not count as observables, a criterion well established in the philosophy of science.

Level 1 of this attribute characterizes statements or renderings that fail or where there is no intention to explicitly describe or predict the real world in terms of observable data, past or present. Evidence in a scientific sense is of no concern at this level. Such a statement might be an original metaphorical construct: fiction, poetry, music, abstract art, or dance. Life would not be the same without metaphor. Feelings and spirituality are represented in abstract words and music, surely connecting to people, but the meaning or essence of the rendering need not represent explicit observable events external to the rendering itself.

Note that for level 1 the rendering *per se* surely consists of observable and measurable properties: music has continuous pitch, painting is done in a spectrum of color, and dance is continuous in motion and time. But in this case, it is up to the observer to make a connection (or not) between the model (metaphor) and the reality. The mapping to the real world is not explicit. Such representations are very different kinds of statements from what are published in scientific papers. We could call them models at a "low" level, because it is not necessarily clear to observers what they are models of. That allows ample room for arbitrary interpretation. Level 1 of the *applicability to observables* category necessarily precludes any such model from being called objective or scientific.

At the other extreme, a level 3 model seeks to predict with precision some future events based on some well-defined independent observable and measurable variables. It represents how the variables interact to produce a result, whether static or dynamic. Level 3 epitomizes what we mean by a model in the book.

Somewhere in between is a description of existing events or objects, a representation of some aspect of the world as it is. It may lack an explicitly stated "IF X, THEN Y" rule to predict new data output from given data input. A photograph is an example, or a desktop model of the Eiffel Tower. Succinct explications of historical events, where there is no effort to generalize regarding lessons for the future, would also fit this intermediate level.

(B)   *Dimensionality*

The attribute of *dimensionality* refers to the number of dimensions of the independent (input) and dependent (output) variable state spaces: how many input and output variables are there in the model. A model can be single-input–single-output, multi-input–single-output, or multi-input–multi-output. (Single-input–multi-output makes no sense since the changes in outputs would be 100% correlated, or are determined by factors not in the model. (For example, increased temperature can increase both volume and pressure, but how much of each depends on other variables.)) The world is complicated, and in general, a combination of many outputs (a vector) is a complex function of many inputs (another vector).

The real world can be said to have an infinite number of variables. To include all inputs and outputs in one gigantic model is totally unrealistic, even for a narrow slice of reality, so the modeler must always put up with variability in the real world that is unaccounted for in the model, presumably by input variables not included in the model. Some such input variables may be known, but their effects unknown. Then there are the unknown–unknowns, variables that the modeler is not even aware of (the so-called "unk–unks"). One might hope that a single output is mostly related to a single input, but that is seldom the case. So, one hopes to find several inputs that relate in combination to a single output, so as to account for most of the variability, that is, most of what accounts for changes in the output variable. Even better (more useful) is to model a relation between a group of inputs and a group of outputs (multiple models in one, so to speak).

Science can only model bits and pieces of reality, and a "model of everything" is not conceivable (though physicists dream of it). A literal model of everything would have to predict the behavior of all elementary particles in the universe and thus every object or event that they constitute.

However, a word about causality is in order. Unfortunately, there is no way to firmly establish independence between variables or causality between inputs and outputs (as distinguished from correlation). To do so would require encompassing the totality of possible variables and their interactions. For example, all changes in output variables could be caused by one unknown factor. Furthermore, when there are closed feedback loops what is considered as input and what as output may be arbitrary. For example, when you push down on a spring, you feel a force—or when you apply a force to a spring, it deflects. Which is the cause, the force or the displacement? Obviously, either force or displacement can be assumed to be the cause, with the other as the effect.

(C)   *Metricity*

The attribute of *metricity* (the property of having measurement) has to do with the meaning of the variables of the model in the sense of the well-known psychophysical scales of S.S. Stevens (1951), namely, nominal, ordinal, interval, and ratio scales. At one extreme, the relationships are expressed in variables that are only nominal (name) categories. At the other limit of metricity are relationships expressed in variables that have a cardinal property of continuous quantitative values (one can consider equal ratio relations as the very limit with equal intervals close behind). In between are ordinals, expressions that X is greater (e.g., better) or lesser (worse) than Y. More specific definitions are as follows:

A *nominal* scale is a naming process. The object or event is assigned a unique number. For a given set of such operations on different objects or events, any change of number assignments to the items is permissible so long as every object or event has its own unique number. Examples are social security numbers, numbers on football player shirts.

An *ordinal* scale assigns numbers to order the objects or events. A greater number means a greater value of some attribute. Any new number assignments are permissible so long as monotonicity (the same ordering) is preserved. Examples are quality ratings of hotels or films, Brinell hardness of surfaces. Sometimes, the order ratings are done by appointed human judges (e.g., professional food tasters), and sometimes by scientific instruments.

An *interval* scale assigns numbers to the objects or events in such a way that equality of numerical intervals is preserved. Any change in number assignments is permissible if it maintains the equality of intervals between any set of objects or events. For example, if *A* and *B* are 2 ft apart and *B* and *C* are 2 ft apart, one can transform the scale to inches or meters and the *A–B* and *B–C* intervals are still the same number (as each other) apart. Note that an arbitrary number can be added to each of A, B, and C and the intervals between them would not change. Temperature measured in degrees Fahrenheit would be an example of an interval scale; zero degrees on this scale has no particular meaning.

A *ratio* scale is an even stricter cardinal scale that assigns numbers to objects or events relative to some absolute zero reference, which means that under any reassignment of numbers to the items, the ratios are preserved. Note that if $A = 4$ and $B = 2$, then $A = 2B$. Transformation to a scale with half-size units would make $A = 8$ and $B = 4$, so A is still $2B$. Note that in this case an arbitrary number cannot be added to each of A and B, for the ratio property would not then be preserved. Temperature measured in degrees Kelvin would be an example of a ratio scale. On the Kelvin scale, the zero has a physical meaning (the molecules stop moving), and a number twice as large as another number is truly twice as hot. On the Centigrade or Fahrenheit scales that would not make sense, since a number twice as large as another number (say 40°C or F) does not mean twice as hot as 20° (C or F) in any physically meaningful way.

Note further that each successive scale (in the order given in the text) retains the properties of the one preceding it. I have lumped the interval and ratio scales into the level 3 metricity category for convenience, though the ratio scale is more extreme than the interval scale.

(D)   *Robustness*

The attribute of *robustness* refers to the breadth of applicability. The least robust model is one that has only one purpose, a unique application, and is otherwise useless. An example might be an instruction for how to operate a particular home appliance or item of software. At the other extreme is a model that has extremely wide applicability, say Newton's law, $F = MA$, that is, force equals mass times acceleration. We can assert that in the limit a model that applies to everything from atomic scale to galactic scale does not exist, though, as noted with the *observables* attribute, physicists are thinking hard about what it would take for a "theory of everything" to exist.

(E)   *Social Penetration*

The attribute of *social penetration* has to do with the degree to which a model is known, understood, accepted, and used by the appropriate community of people. The scale goes from purely mental models that are confined to internal thoughts of some one person to models that are widely understood, accepted, and applied by a large and diverse community of people. In between are models that are described to others or published in the literature, perhaps in competition with other models for the same application, and perhaps used by only a few practitioners.

Social penetration is perhaps the most difficult attribute to achieve, or at least the most time-consuming aspect of modeling. Any new model, particularly if it threatens the old, will be challenged (not a bad thing, science depends on it). Opponents will question the observations, as well as its relevance in terms of addressing a salient question. In a sense, a new model must be actively marketed by its advocates, or else it may never be accepted. However, success in marketing a model (having more people believe it and use it) does not by itself mean it is more useful and reliable. The model can be fraudulent, in which case its legitimacy will eventually be questioned as more evidence is gathered. For a very long time, most people believed the earth to be flat. If a model is marketed as answering some question, one can also challenge whether the question is meaningful, for example, what is the taste of red.

(F) *Conciseness*

The attribute of *conciseness* is added to the taxonomy to provide a metric on brevity of presentation, along with adherence to denotation, clarity, and reason. Adherence to brevity, denotation, clarity, and reason is the principle that has come to be called "Ockham's razor." This principle was originally attributed to philosopher William of Ockham (1495), according to his pronouncement *Numquam ponenda est pluralitas sine necessitate* (plurality must never be posited without necessity). The idea is that one should use the simplest statement that does not compromise explanatory power. Of course, too few words may reduce explanatory power.

A hundred-page exposition of an argument, theory, premise, or relationship would not normally be considered as a model. This book is not a model, though this taxonomy of model attributes is a model.

EXAMPLES OF MODELS IN TERMS OF THE ATTRIBUTES

It may be helpful to consider examples of models that combine different levels of the attributes. Note that any one model can have different levels of any given attribute. Consider the lettered rows and numbered columns of Table 2.1. We assume that more of the attribute of conciseness always makes a better scientific model, other things equal, so for simplicity that attribute is excluded from the various combinations compared in the following text.

[A1,B1,C1,D1,E1] is a mental model (a musing) assigning some hypothetical simple relation between hypothetical entities, for example, a particular imagined unicorn is green.

[A1,B1,C3,D3,E1] is a mental model on a provided numerical or category scale. By itself, it cannot be called an objective or scientific model. Only when it is aggregated with other such responses, all of which can be said to be responses on the same scale to a common stimulus, can it begin to have the makings of such a model.

[A2,B2,C2,D2,E2] is a hypothetical framework of relationships with moderate focus communicated to an interested community. The present taxonomy of models might be an example. By itself, it is only an expression of my idea in my language, but it can become accepted as an objective (though qualitative) model if others agree.

[A1,B2,C2,D1,E1] is a mental model (contemplation) by an experimenter to rank order particular experimental data for later analysis (and an eventual model).

[A1,B2,C3,D1,E1] is a mental model based on detailed numerical relationships between known data that are broadly applicable within some special industrial context. This might be considered a way to analyze cardinal data (from observations) in a given experiment and lead to a model.

[A2,B1,C3,D2,E3] is a widely accepted quantitative depiction of past or present facts such as a graphical plot of past stock prices.

[A3,B2,C2,D2,E3] is a predictive, ordinal, well-accepted model of established data with moderate applicability. This might be an experimentally based qualitative or best-practices design guideline that is standard within a given industry.

[A3,B2,C3,D3,E3] is typical of widely accepted and applied models in science and engineering such as the physical laws of Newton, Ohm, Bernoulli, and Faraday.

At the extreme limit, [A3,B3,C3,D3,E3] is a fully predictive, quantitative model that applies to and explains all of nature and is accepted without hesitation. It is a presentation of relationships between well-defined variables pertaining to all objects or events, communicated to others in a combination of written words, symbols, or graphic images that conveys quantitative structure including explicit cardinal relationships and is demonstrably communicated widely and understood and accepted for use. It is based on data that are empirically observable by anyone, has a demonstrable record of being predictive within a wide set of circumstances and with high statistical confidence, does not conflict with other models of relationships between the same variables, and applies at all magnitude levels of nature from subatomic to intergalactic. This is what physicists aspire to as the previously mentioned "theory of everything" but have not yet reached, and surely never will.

WHY MAKE THE EFFORT TO MODEL?

Modeling takes effort. To find the right words, words that have common meanings; to say what needs to be said and say it succinctly; to get the diagram, graph, or other image just right so that it communicates; to do the math correctly if the math is appropriate to make the case—all of this takes real effort. Why do it? What's the point?

One instigates a modeling effort for the same reason that one investigates an interesting plant on the hiking trail, makes a note in one's diary of a special personal interaction, or follows up any observation that arouses interest and curiosity. Modeling engenders satisfaction, better ability to share the experience with other people, and the improved likelihood of making some predictions that might be useful later on.

Though individuals can gain insights in various ways, for the scientist modeling is the *sine qua non* for staking public claim to some new insight, or asserting a better way of considering some aspect of the world and communicating it to colleagues. From a crass perspective, a new model might help get a paper published.

But there is a further, probably more important reason. That is, the very process of modeling forces one to think hard about the slice of nature under consideration, to ask and answer the question of what are the essential features of the structure and function, and to make "If X, then Y" predictions. Committing to a model is tantamount to committing to think hard. It involves putting one's reputation on the line, which surely motivates the thinking process.

ATTRIBUTE CONSIDERATIONS IN MAKING MODELS USEFUL

Useful models, where *useful* is broadly defined, span the full range of the taxonomy, short of the physicist's ultimate "theory of everything." The impression must not be given that useful models are easy to come by, that they always tell the truth (complete truth is never known), and indeed that more of each attribute necessarily makes for a more useful model (too much math will make it incomprehensible to most people). For those reasons, it is important to add some caveats.

With regard to *applicability to observables*, a model builder at first may consider, at the A1 level, verbal propositions or hypothetical frameworks that are not bound by any specific past observations (e.g., hypotheses and conjecture about the problem at hand, how to bound it in terms of what independent and dependent variables to include, what is known and what is unknown, and what is important). As the model develops, explicit data from completed experiments and experience should be applied at the A2 level. The goal is the ability to predict future observable data or physical system measurements (A3).

In an effort to "sell" the model to colleagues, modelers have been known to fudge the data, or fake it completely, or select those observables that support the desired conclusion and ignore the others. We see that in political campaigns, where protagonists for one side or the other point to models based on carefully selected observables that support their position, and carefully avoid including observables that support the other side. Alternatively, once the model is complete and makes a projection, there is a tendency to interpret the result in a manner to support the modeler's bias.

High levels of the *dimensionality* attribute are common in models that are inherently complex with many variables (i.e., those in social, economic, and medical fields). When quantitative measures are available modelers may tend toward factorial experimental designs that cater to analysis of variance. But more often, multi-input–multi-output relationships tend to be those used in more qualitative (low level of metricity) modeling. Such models are employed by clinicians (e.g., a particular pattern of maladies is associated with a given pattern of symptoms), where the intent is to check for obvious effects of a drug or some policy intervention.

One pitfall of modeling is encountered when building a descriptive model based on a given set of data by using a large number of model parameters. The model may then result in a good fit to the particular data. But then if the same set of input variables has slightly different values, then the model may not fit the new data at all; in other words, the model may have no predictive capability. A standard joke is that with enough parameters, one can concoct a model that will draw an elephant or whistle Dixie. This problem is characteristic of models based on a large set of rules, for example, one rule for each variable, where each rule is more or less tailored to the specific value of the variables in the original data set.

The *metricity* attribute is mostly a matter of the modeler's intent. Human behavior can be represented in words (C1), and much social science literature does that and no more. Human performance modeling mostly implies at least ordinality (C2): that something is bigger, faster, or better by some criterion. Models that predict in cardinal relationships (C3) are most desirable, because cardinal (continuous numerical) relations subsume order relations but offer greater precision.

But there are dangers. Models that predict numbers and involve complex equations may appear to be more sophisticated (and more easily get published in scientific journals); but for many applications, they may be of less use than a very simple model in the nominal or ordinal category that is easier for users to understand.

The *robustness* attribute is partially a matter of intent. Any reasonable modeler knows that the applicability of any human performance model is initially quite limited (D1), though the modeler might hope that the model can gain wider applicability with time and revision to include other variables and wider use. Being human, the modeler is naturally a protagonist, and may easily get caught up in marketing the model beyond what is warranted.

In the development of any model, the *social penetration* evolution begins with the initiator's mental model (E1) and possibly evolves over time to full acceptance (E3) by the community of experts and practitioners in the field. Usually, this does not occur in a linear, one-directional process. Typically, a scientific model becomes open to review and criticism by peers in the submission/publication process of any reputable journal. It may be rejected at first, but with suggestions to the author and rethinking, the model may then be refined. There will always be a delay in publication. After initial application, it may be subjected to further active feedback and refinement to make it more understandable and useful.

Adhering to the *conciseness* attribute requires effort on the part of the model developer. That effort pays off in making the model more easily reproducible (it saves space) and transferable. That also makes it more useful.

One of the highest quality (in the sense of this taxonomy) human performance models ever developed is the McRuer and Jex (1967) model that specifies the explicit human transfer function (as a first-order differential equation plus time delay) to characterize the combined human operator and controlled element in a target tracking feedback system. It is elegant because it has a fixed form and comes with a table of (relatively few) parameters with specified dependence on the input bandwidth and controlled element dynamics. In that sense, it conforms well to the edict of Ockham's razor (simpler is better). This is an [A3,B2,C3,D2,E3] model. It evolved after many years of government funding to many investigators (including this author) interested in pilots flying high-performance fighter aircraft and maintaining control stability. However, even though the McRuer–Jex model set a high standard for human performance models, it is of relatively little interest today because now aircraft are mostly flown on autopilot or fly-by-wire software that provides automated compensation to prevent control instability.

In a different vein, a modeling approach called Adaptive Character of Thought—ACT-R (Anderson, 1990) represents procedural knowledge in units called *production rules*, interacting with declarative knowledge called *chunks*—all implemented in computer software. This approach is now widely used by other behavioral modelers for car driving, for example. The general approach aspires to the [A3,B3,C3,D2,E3] category. However, one problem with such complex simulator models is that they contain many parameters that are adjusted to provide fit to the available data for the given application, so that predictability to other situations is limited.

Models are now being used in an astonishing new way. A model of some technical function couched in computer logic (e.g., for an aircraft control system) can be made to directly specify the software code without a programmer ever having to write specific and detailed lines of code. The danger herein is that no human may ever know exactly what instructions the computer has been given. This is the sort of problem that the "father of cybernetics" Norbert Wiener warned about in his 1964 Pulitzer Prize-winning *God and Golem Inc.*, that computers may do things they were programmed to do, but those actions were never really intended by the human programmer.

One final caveat regarding usefulness of models: while it may seem that the explicit prediction (for a given input what is the output) is the most useful property of a model, that is often of only secondary importance. The most useful aspect may well be that the conception, development, and publication of the model caused people to think hard about the problem—what are the variables that count the most, how best to formulate the problem, and so on. A model requires the model builder to think, and it should also require the model reader/user to think, and more thinking is usually a good thing.

SOCIAL CHOICE

The aggregate of our publicly accepted models forms our store of what we reasonably assume we know. Individual mental models encapsulate individual subjective beliefs, whether based on evidence or based on faith, but they do not contribute to the general store of knowledge—they are private. Unfortunately at this point in human evolution, we mostly lack objective means to capture individual mental models and to compare them or combine them where they agree. Of course, people can share mental models by verbal expression or by acting on their beliefs, as in coordinated athletic team sports, where individuals infer each other's mental models as they interact.

One objective means to discover a mental model is using voting to capture the mental preferences or beliefs of a group. Pollsters work hard in posing simple questions to get valid answers about what people really believe. The problem here is that the simple questions are posed in words that often have ambiguous interpretation: Do you believe in X? What kind of X? Do people all mean the same thing by what is being called X? What does "believe" mean? Do I believe X or not believe X because that is my tradition, that is what is comfortable to say I believe? Did I really ever question my belief in X?

We tend to make key decisions in groups and communities by voting. Group decision-making is called *social choice*. Economists who have studied the issues of social choice reveal certain difficulties in finding the preferences of a group. Kenneth Arrow (1950) won the Nobel Prize for, among other things, showing that it is logically impossible for people offered choice among at least three alternatives to always come up with a clear choice by majority vote. Of course, there can be tie votes. If you allow people to rank-order preferences among A, B, and C, one person can prefer A to B to C, another can prefer B to C to A, and still another C to A to B. There is then no group preference. The latter is called *preference intransitivity*. It can get much more complicated. It is probably just as well, because, while one might like to establish public policies such as those that impinge on religious belief according to majority voting, tyrannizing the minority opinion on religious belief seems particularly inappropriate in our culture

WHAT MODELS ARE NOT

There is a danger that models can come to be regarded as algorithms for doing research. They are mental handles for thinking about aspects of the world, in this book limited to human–system interaction. They are *not* algorithms for initiating a research project, for selecting some phenomenon to be investigated. They are *not* guides for experimental design, though they may give clues to what to select as independent and dependent variables, a critical discipline that involves knowingly omitting some variables even though they may be measured and/or controlled. They do *not* determine what kind of controls to use, or the degree to which the experimenter should be present and/or involved in administering the independent variables (stimuli to the human subject or system). They do *not*, in most cases, dictate whether measurements should be cardinal, ordinal, or nominal and by what means the data are collected. In general, the model should emerge, after the fact, from the experimental design and the results, rather than the other way around.

Some lay people express dismay that behavioral scientists even contemplate "putting people in boxes" as being an inhumane undertaking that cannot possibly capture the rich and subtle nature of human beings. Of course, they are correct that science can never reduce any single person or group of humans to a mathematical equation or a diagram, but neither can medicine or economics or any other mode of describing a person do that. In doing science, one is always limited to a very few independent and dependent variables in the face of variability, which is really a way of expressing residual ignorance about why the data turned out as it did. Indeed the variability, the representation of what is beyond the model, is a critical consideration in publishing any experimental results.

3

IMPORTANT DISTINCTIONS IN MODELING

Beyond the six model attributes detailed before, there are several other distinctions that are important to consider.

OBJECTIVE AND SUBJECTIVE MODELS

Necessarily, a model is a representation of very limited aspects of the thing or events being modeled. In that sense, it can be said that *all models are wrong* with respect to the full reality of the slice of nature being modeled. Consider a global map of the world. The globe is not the same as the real world. The globe is a very different size. The globe has different colors to identify countries. The real world is not colored the same as the globe. The globe shows distances between cities, but the distances are not the same as those of the world. What is the same are the *relative distances* between cities and the *proportional spatial relations* of rivers and country boundaries. That is all. But as such, this model is very useful and aids understanding with regard to those particular attributes.

Increasingly, science and technology and government and industry are being driven by models. In physics, for example, our understanding of the

Modeling Human–System Interaction: Philosophical and Methodological Considerations, with Examples, First Edition. Thomas B. Sheridan.
© 2017 John Wiley & Sons, Inc. Published 2017 by John Wiley & Sons, Inc.

universe is largely based on model extrapolations well beyond what we can observe directly, and huge experimental efforts are made to verify the models (e.g., the hunt by particle physicists for the Higgs boson). Social science is definitely progressing, and in the future may well be aided by progress in neuroscience, but has not come close to that level of sophistication in quantitative modeling as now used in physical science and engineering.

For all fields of science and technology, models serve the function of asserting in a public way what the modeler believes to be true, thereby allowing for criticism and refinement by the relevant community. It may be said that to the extent that we can model, we have a basis to form consensus and therefore have useful knowledge. In this sense, the usefulness of the model is in forming a belief system about the domain of interest.

As noted earlier, a scientific model is fully denotative and rational, as contrasted to one that is essentially connotative, such as a novel, a poem, a myth, or an artistic statement (abstract or nonrealistic painting, music, or dance). A scientific model adheres to strictures of the scientific method (explicated earlier in Chapter 1, Section "Scientific Method: What Is It?"), such as having a basis in objects or events that are observed by two or more people (and potentially observable by anyone). It is stated concisely. Often it aspires to generality, implying (or explicitly stating) that the model applies not only to the objects or events observed but also to things other than the ones used as the basis of the model. Such generality is achieved in science by stating that certain dependent (output) variables are specific functions of certain independent (input) variables, where both sets of variables are well defined and observable to anyone equipped with the required means to measure them.

The quality of any model, but especially a scientific model, depends on the number of variables that can be accommodated simultaneously, the rigor of the measurement process that goes with the variables, the robustness or breadth of applicability of the model, and the degree to which the model is understood, accepted, and applied by the peer community.

One can cite plenty of examples of where scientific models have failed. Failure typically occurs when there is a rush to apply modeling where it has not been previously tried, and the target problem has not been thought through sufficiently, or the modeler expects too much too fast. An outstanding example is the Wall Street debacle of 2008, where the "quants" employed models to take statistical risks that were much greater than what was warranted, and the result was not pretty. More recently, the bond rating agency Standard and Poor was sued by the U.S. government for basing ratings on a proprietary model that was known to be faulty, an action that also helped precipitate the financial crisis of the same period. Normally, models are published in the open literature and are peer-reviewed, but it is a right of any institution to maintain privacy. They do so at their own risk.

The term *simulation* is often used in conjunction with modeling. Simulation, today usually referring to simulation by computer, simply means putting the logic and mathematics in computer software form and then running computer trials to test what different inputs to the model produce as outputs. Computer software is a special form of scientific modeling.

Computer scientists like to use the philosopher's term *ontology*, which in their domain is defined as meaning "formal representation of knowledge as a set of concepts within a domain, and the relationships between those concepts." (In philosophy, ontology means a *theory of being or existence.*) The computer community emphasizes that an ontology is a "specification of a conceptualization," where the stated purpose of designing ontologies is to share knowledge and make intellectual commitments. As a practical matter, computer scientists are compelled by these formalities to make the various bits of software (they call them *agents*) work together. Thus, a common ontology defines the vocabulary with which queries and assertions can be exchanged. Ontological commitments are agreements to use the shared vocabulary in a coherent and consistent manner. The computerized agents sharing a vocabulary need not have the same knowledge (share the same knowledge base). Each may know things the others do not, and is not required to answer all queries that can be formulated in the shared vocabulary. In short, a commitment to a common ontology is a guarantee of consistency, but not a guarantee of completeness. That is a reasonable way to think about models in general, and how people can use models to accomplish useful goals. (In any case, the formality makes the computer scientists sound like they know what they are talking about!)

SIMPLE AND COMPLEX MODELS

A simplest model is a statement that X is Y, where X is a thing or event and Y is a descriptor (adjective, number, etc.). A very complex model might be the complete engineering specification of a jumbo jet aircraft or the voluminous U.S. Code of Federal Regulations. In the simplest model case, the independent variables have to do with what qualifies as X. In the case of the jumbo jet, the independent variables have to do with what exact part of the airplane and what function of that part we are talking about. For the Code of Federal Regulations, the independent variables would provide information enabling a user to go to a specific chapter and rule and determine what the regulation stated. Some readers might complain that I am using the term "model" too broadly here, but there is really no way to restrict the term to a narrower range.

DESCRIPTIVE AND PRESCRIPTIVE (NORMATIVE) MODELS

A *descriptive* model serves to tell how some specific thing is structured or to provide detail on some set of events. It looks back in time in the sense that it refers to objects or events that exist or have existed in the past. If the model describes an object yet to be built or a plan yet to be implemented, the descriptive model is based on the already completed design or plan. In contrast, a *prescriptive* or *normative* model tells what *should* happen in the future according to specified or assumed norms. They can be social/cultural norms or they can be physical norms (e.g., how some aspect of the world works: that a (given) input should produce a stated output). With luck, a descriptive model is prescriptive for a new situation that is similar to that on which the model was based.

STATIC AND DYNAMIC MODELS

In a *static* model, neither the input nor the output variables change with time, whereas in a *dynamic* model either input or output or both vary with time. For example, when you press down with force F on a spring having stiffness parameter K it will deflect a certain distance X. That is a static model; nothing changes with time. If a clock pendulum is initially displaced from a resting position to an angle A and then released, a simple dynamic model will predict the trajectory of how it will swing back and forth with time as a function of initial angle A, the pendulum length and the force of gravity. An economic dynamic model might have as input the continually changing price of gasoline over a 1-year period and output the modeled effect of those changes on vehicle use over the same period. Static models require only algebraic equations, whereas dynamic models make use of differential equations with time as the argument. Thus the output of a dynamic model is a curve that plots one or more variables against time.

DETERMINISTIC AND PROBABILISTIC MODELS

A *deterministic* model says that if X is the input, then Y is the output—for certain. A *probabilistic* model says that if X is the input, then Y is the output with some probability less than 1. More generally for input X, the probabilistic model will specify a set of outputs each with a different probability (possibly a probability distribution). A common type of probabilistic model (called a *Markov* model) is a tree-graph that branches from one or more nodes representing initial states, where the branches to downstream or peripheral nodes are tagged with probabilities. Starting from any initial node, the probabilities of ending in any downstream node are thus calculable.

Whether the model is static or dynamic, a computer can be programmed to try a large number of different inputs to test the resulting outputs. The inputs can systematically proceed by intervals through a given range of values, or can be selected randomly from a given distribution. The latter is called a Monte Carlo model after the casino gambling reputation of the Principality of Monaco.

HIERARCHY OF ABSTRACTION

Rasmussen et al. (1994) have made an important and useful description of the functional properties of a system according to what they call "goals–means" or "means–ends relationships" in a functional abstraction hierarchy. What follows is a description (from Wikipedia).

Functional Purpose The functional purpose level describes the goals and purposes of the system. This level typically includes more than one system goal such that the goals conflict or complement each other. The relationships between the goals indicate potential trade-offs and constraints within the work domain of the system. For example, the goals of a refrigerator might be to cool food to a certain temperature while using a minimal amount of electricity.

Abstract Function The abstract function level describes the underlying laws and principles that govern the goals of the system. These may be empirical laws in a physical system, judicial laws in a social system, or even economic principles in a commercial system. In general, the laws and principles focus on things that need to be conserved or that flow through the system such as mass. The operation of the refrigerator as a heat pump is governed by the second law of thermodynamics.

Generalized Function The generalized function level explains the processes involved in the laws and principles found at the abstract function level, that is, how each abstract function is achieved. Causal relationships exist between the elements found at the generalized function level. The refrigeration cycle in a refrigerator involves pumping heat from an area of low temperature (source) into an area of higher temperature (sink).

Physical Function The physical function level reveals the physical components or equipment associated with the processes identified at the generalized function level. The capabilities and limitations of the components such as maximum capacity are also usually noted. A refrigerator may consist of heat exchange pipes and a gas compressor that can exert a certain maximum pressure on the cooling medium.

Physical Form The physical form level describes the condition, location, and physical appearance of the components shown at the physical function level. In the refrigerator example, the heat exchange pipes and the gas compressor are described as being arranged in a specific manner, basically illustrating the location of the components. Physical characteristics may include things such as color, dimensions, and shape.

"Such a hierarchy," says Rasmussen, "describes bottom-up what components and functions can be used for, how they may serve higher level purposes, and, top-down, how purposes can be implemented by functions and components. During system design and supervisory control, the description of a physical system will typically be varied in at least two ways. The description can be varied independently along the abstract–concrete dimension, representing means–ends relationships, and the dimension representing whole–parts relationships."

SOME PHILOSOPHICAL PERSPECTIVES

In the preceding sections, we referred to models as concise, logical representations of belief that could be called scientific, many adaptable to quantification. This section offers diverse perspectives on what it means to believe what is true. These are modeling perspectives with a softer edge that tend to be qualitative, and are more debatable and without consensus among scholars.

Initially, the user of a model generated by someone else cannot understand the full implications of what the modeler intended. The truth cannot be absorbed in one bite. Belief evolves. Nor can the modeler appreciate where that reader is coming from, what life experience that reader has that will affect their initial interpretation. Further, no matter how sophisticated, how elaborate the model, the whole truth will never be told: there is just too much to tell; there are just too many variables to the reality of the whole truth. Finally, the modeler will never know the whole truth of whatever the model is about. Much as any modeler would aspire to tell "all the truth," that is clearly not possible: it ends up being "slant," as in the famous Emily Dickinson poem (Dickinson, 1924). We noted earlier the common expression that "all models are wrong." Modelers who are wedded to a particular model and are overly aggressive in marketing it are sometimes disparaged as claiming that they "Have model, will travel." Dickinson's "gradual dazzling" of the truth requires feedback, clarification, further communication, and mental soaking time for both modeler and reader. An established model, one with a high level of social penetration, represents a convergence based on gradual refinement and necessarily on the passing of time.

Philosophers, logicians, economists, and others too numerous to review have been proffering theories of belief for decades. From the ancient Greek philosophers until today, there have been lively philosophical arguments about whether we believe based on rationality or on emotion. For example, philosopher David Hume (1739) claimed that "Reason is, and ought only to be the slave of the passions." Allegedly, he was referring to one's innate sense of hopes, wishes, and morality, and that in dealing with the issues of life one cannot reason independently of one's passions. We know that emotion plays a large role in belief, whether in a moral context or not.

Charles Sanders Peirce (2001) characterized inquiry not so much as an effort to gain truth but as a means to avoid doubts, social disagreement, and irritation, so as to reach a belief on the basis of which one is prepared to act. He distinguished several approaches to inquiry such as sticking to some original belief in order to bring comfort and decisiveness (he called this *tenacity*), being brutally authoritarian and trying to dominate others who offer contrary evidence (he called this *authority*), and conformity with current paradigms, taste, and fashion, that is, what is more respectable (he called this *congruity*), and finally the method of science, which seeks to criticize, correct, and improve upon itself. I suppose real human inquiry is a blend of all of these modes.

G.E. Moore (Levy, 1981) posed the following paradox, namely that people are inclined to say things such as "It is raining but I can't believe it is raining," which is patently absurd but nevertheless logically consistent. This led Moore to assert that "one cannot believe falsely, though one can speak of someone else that they believe falsely."

Modern economic and cognitive science has worked hard to formalize the models of belief. It continues to be recognized that belief is not simply a matter of being rational and having a straightforward way to draw conclusions about what is true and what is not. If one knows the probabilities associated with various kinds of observable evidence (probability of a particular observation given that a particular hypothesis is true), there is a well-known mathematical model called Bayes' theorem that works well to estimate the truth or falsity of premises (see Chapter 4, Section "Symbolic Statements and Statistical Inference"). But the problem is that those contingent probabilities are subjective and may be only vaguely held in mind. This goes for both the mental formulation of the hypothesis and the estimation of probability contingent on each hypothesis. Further, some hypotheses may be well formed mentally along with some basis for estimating their probabilities, whereas some other plausible hypothesis may be overlooked or offer no hint of its probability. In using a Bayesian approach a "don't know" category is often treated as indifference with 50–50 probability belief. But some belief analysts rebel against interpreting "don't know" as a 50–50 probability estimation, and regard that judgment as very different from making a probability assessment.

There has been recent formal (logical, mathematical) work on belief called *Dempster–Shafer belief theory* (Shafer, 1976). In some sense, it is a generalization of Bayes' theorem cited before as one of the example models concerning statistical inference from evidence. It differs from Bayes' theorem, in that the procedure is not limited to subjective probability judgments directly on the proposition. Rather, it is based on deriving degrees on belief for a given proposition from subjective judgments for both *strength of belief* and *judgment of plausibility* regarding various other questions (sets of *possibilities*) that contain the proposition. For example, a set of four possibilities might be as follows: (1) X is true, (2) X is false, (3) X is neither true nor false, and (4) X is either true or false. There is a rule in Dempster–Shafer theory for combining multiple degrees of belief when they are based on elements of evidence that are mutually independent. The degree of belief in the main proposition depends on the answers to related questions and the subjective probability of each answer. There is the further problem of *unknown unknowns* (unk–unks). Some relevant properties are known as credible variables, that is, previously observed and identified as variables, their values unknown. Other properties have never been known or considered: the unk–unks. The believer or modeler simply has no thought of their existence or their relation to the issue at hand. This suggests that formal belief models should include two components: (1) the current state of belief and (2) a rule for how current belief would be updated in light of new variables and new evidence. Unfortunately, the second component is missing in most cases.

Trust is a topic (Chapter 6, Section "Trust") that has been left outside of science until fairly recently when economists and psychologists picked it up. A new Journal called *Trust Research* appeared in 2011. System engineers concerned with safety have especially become interested in trust. The topic of virtual reality is also discussed later (Chapter 5, Section "Experiencing What Is Virtual: New Demands for Human–System Modeling") because it offers means to conduct experiments on human subjects who can be shown to experience (and believe in) phenomena that physically do not exist. Finally, I include a model I particularly like that suggests how knowledge evolves (Chapter 8, Section "Control by Continuously Updating an Internal Model"). It is a model commonly used by engineers to make computers learn.

4

FORMS OF REPRESENTATION

VERBAL MODELS

Most of what can be called models are expressed in words only. Historians, philosophers, theologians, newspaper columnists, sportswriters all use words in print media to express their beliefs about how some things or events are structured or relate to other things or events. Some of these can be called scientific models because they explicate concisely the structure or function of some things or events. In the most general case, novelists, poets, and lyricists are verbal connotative modelers, but we choose to exclude those models because they do not intend to denote the explicit truth.

In the following, for example, are Charles Darwin's (1859) own words in *On the Origin of Species* that I have taken to be a summary explanation of his theory of natural selection, more commonly known as the theory of evolution. Of course, he wrote many more words of explanation in that famous work. But these words do exemplify what could be called a verbal scientific model.

> Evolution by natural selection is a process that is inferred from three facts about populations: 1) more offspring are produced than can possibly survive; 2) traits vary among individuals, leading to differential rates of survival and reproduction; and 3) trait

Modeling Human–System Interaction: Philosophical and Methodological Considerations, with Examples, First Edition. Thomas B. Sheridan.
© 2017 John Wiley & Sons, Inc. Published 2017 by John Wiley & Sons, Inc.

differences are heritable. Thus, when members of a population die they are replaced by the progeny of parents that were better adapted to survive and reproduce in the environment in which natural selection took place. This process creates and preserves traits that are seemingly fitted for the functional roles they perform.

Darwin's theory is often stated as "survival of the fittest." The phrase is clearly not an adequate verbal model since it says nothing about the mechanism by which traits are passed on to the later generations of the same species. Any model must have sufficient explanatory detail.

Essentially, all scientific models in their presentation in scientific papers have to be augmented by words, necessary to explain what the model is all about, state the limitations, and help the reader navigate the graph, diagram, or math. That does not necessarily make such papers verbal models.

GRAPHS

Many models render their message mostly in terms of graphics, usually accompanied by a few words to identify the variables and assist in explanation. The graphics can be line graphs, charts, diagrams, maps, tables, or photos. Some pictures really are worth a thousand words, and need little additional explanation. Some kinds of explanations are very difficult or impossible to convey in words, and a graphic is essential. Figure 4.1 is an example. It is a model that tells the user about quantitative relationships in the given context.

Plots of data fitted by mathematical functions are a common form of model. For example, consider the well-known bell curve, also known as the Gaussian probability density function (Figure 4.2). For whatever context this

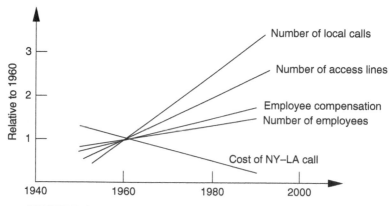

FIGURE 4.1 Trends in telephone company data (hypothetical).

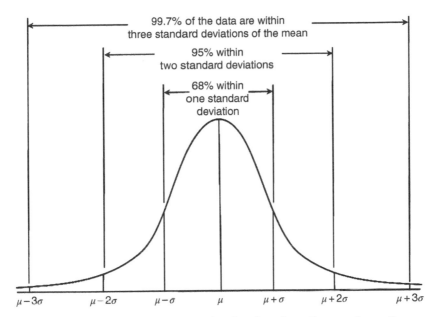

FIGURE 4.2 Gaussian probability density function. Source: https://commons. wikimedia.org/wiki/File:Empirical_Rule.PNG. Used under CC BY-SA 4.0 https:// creativecommons.org/licenses/by-sa/4.0/deed.en.

image is used, it usually tells the viewer that, given some value on the x-axis, the value shown on the y-axis is (or is an estimate of) its probability (likelihood of occurrence relative to other values). This curve is commonly used as a model because of its attractive mathematical properties, namely that there is a precise mathematical function relating probability P to x. Observed data may or may not fit the model, and never will fit exactly.

There are many different mathematical functions that are used for fitting empirical data. If one is lucky enough to find a good fit, that mathematical function becomes an easy way of generalizing the data. But there is nothing magical about any particular form of abstraction (the math) as related to the reality (the data). Whatever mathematical function fits the data is as good as any other, except that some may be simpler to use.

Often a relationship between two or more functional plots is what comprises the model. Figure 4.3 shows the generic form of supply–demand curves that are basic to economics. As the selling price of a given product goes down, there will be more demand by buyers (demand curve). From the producer's perspective, as the price increases there is greater profit and therefore greater willingness to produce (supply curve). At the crossover point, there is an equilibrium that determines what will tend to occur in practice.

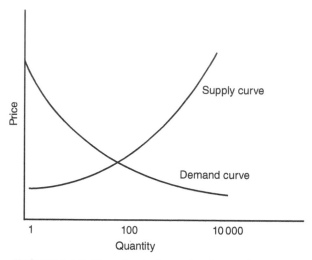

FIGURE 4.3 Hypothetical supply–demand curves.

FIGURE 4.4 Map of the United States. Source: https://commons.wikimedia.org/
wiki/File:Map_of_USA_States_with_names_white.svg. Used under CC BY-SA 4.0
https://creativecommons.org/licenses/by-sa/4.0/.

MAPS

An ordinary map is another kind of graphic model that normally needs words
only to identify locations on the map or identify symbols. The main message
is the spatial layout, the graphical relation of the mapped entities to one
another. Figure 4.4 is an example.

SCHEMATIC DIAGRAMS

Engineers and many scientists like diagrams with arrows, indicating causality relationships. The blocks in schematic diagrams can represent quantitative functions or qualitative relationships.

One of the best-known schematics in the cognitive engineering world is Rasmussen's (1983) diagram (Figure 4.5) of the three levels of behavior: skill-based behavior (learned skills that are executed with little or no conscious mental activity); rule-based behavior (learned or read from instructions, specifying IF X, DO Y), and knowledge-based behavior (meaning a person must think through a situation *ab initio* and draw from memory to devise an appropriate action).

Another favorite use of schematics is for showing taxonomies or comparisons between elements of a situation, sometimes in two or more dimensions. Wickens' diagram of human multiple resources (sensory modalities on one axis, response modes on another axis, and the stages of cognitive processing on a third) is a popular and much cited way to look at human capabilities in relation to one another (Figure 4.6). The stages are modified to correspond to the four stages used in this book: acquire information, analyze information, decide on action, and implement action.

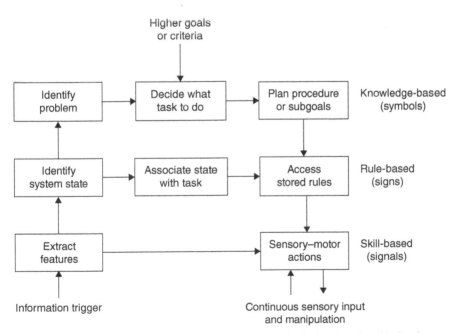

FIGURE 4.5 Rasmussen's schematic diagram depicting levels of behavior.

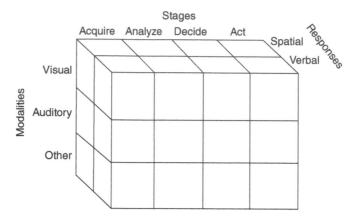

FIGURE 4.6 Wickens' (1984) model of human multiple resources (modified by author).

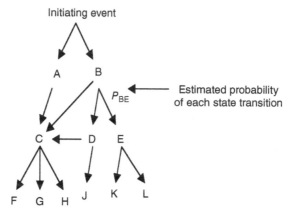

FIGURE 4.7 Forward chaining tree.

LOGIC DIAGRAMS

Logic tree models display component elements and tell the user of a relationship between the elements in causation sequence. For example, Figure 4.7 depicts a forward chaining tree, commonly also called an "event tree," where events are labeled by letters A, B, C,.... It shows that following some initiating event (arrows indicate time sequence) either A or B can happen.

In a forward chaining tree, transition probabilities are assigned to the lines connecting the initiating event to A and B or any other pair of letters (P_{BE} is shown for the B to E transition). For example, assume the probability from initiation I to A is $P_{IA} = 0.7$, and from initiation to B is $P_{IB} = 0.3$. The two numbers must add to 1 since the model implies that there are no other

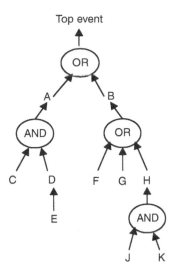

FIGURE 4.8 Backward chaining tree, where AND indicates necessity and OR indicates sufficiency.

possibilities. Now C is shown to happen because of either A or B, and given that B happens C, D, or E can happen with some transition probabilities (that again must add to one). The probability of any event occurring is equal to the probability of the previous event multiplied by the transition probability to the new event (e.g., the probability of E occurring is P_{BE} times 0.3). Working our way down the graph, one can calculate the probabilities of occurrence of each final state F, G, H, J, K, or L.

This model is useful for representing the behavior of any complex system, including people or teams of people. Where do the probability numbers come from? They are derived from recorded frequencies of events that actually happened. With enough frequency data, one has a pretty good estimate of the transition probabilities. Event tree models are particularly useful in analyzing the ways that various accidents can happen.

Another type of causality model may be represented in what is often called a backward chaining tree (Figure 4.8). This type of network model combines the logic of necessity and that of sufficiency. It is often used by safety analysts in representing the chain of causation in accidents, and hence is referred to as a fault tree. It is probably the most reasonable single technique to use in modeling accident causation. The logical relationships are indicated through symbols of AND and OR logic, where the ANDing of two or more events means that both those events are necessary for an indicated result to occur, while the ORing of two or more events means that either event is sufficient for the next event to occur. One starts from the top event and considers what possible events or combinations of events might cause that event to occur. The diagram indicates that either A or B is judged sufficient to have caused the top event. Both C and D are modeled as necessary to cause A, while any of F, G, and H are sufficient to cause B. Both J and K are modeled as necessary to cause H. Developing this type of model is tedious, but is especially useful as a discipline of thinking about what causes what, including what is necessary and what is sufficient. It is commonly used in analyzing causality chains in accidents or spread of diseases.

Graphical models can get very elegant. The physicist Richard Feynman developed what are called Feynman diagrams to portray on a space versus time plot the interactions of subatomic physical particles. They will not be described here because they entail significant understanding of particle physics. For entertainment, a reader interested in graphic complexity might look up the term *complexity* in a Web search and find a truly stunning array of graphical images.

CRISP VERSUS FUZZY LOGIC (SEE ALSO APPENDIX, SECTION "MATHEMATICS OF FUZZY LOGIC")

A new analytical tool has found increasing acceptance as a way to represent the "soft associations" exhibited by the overlaps of meaning between words. It is called *fuzzy logic* or *fuzzy set theory* (Zadeh, 1965). It has applicability to making decisions involving a large number of variables where the many rules available for deciding on an action in any specific context are only available as verbal statements (which is the case for the explication of most real knowledge in this world). An example of applying fuzzy logic is found in Appendix, Section "Mathematics of Fuzzy Logic."

Fuzzy logic has a fascinating history of rejection by computer scientists in the West, whose crisp mathematical ways of thinking have proven ineffectual in dealing with soft, overlapping ideas, and by technology managers who are put off by the term *fuzzy* as being unscientific. At the same time, theorists in Asia have pushed ahead not only with fuzzy theory but also with applications. Systems based on fuzzy logic are now in digital cameras, washing machines, subway speed controls, and automobile transmissions.

Fuzzy logic allows for evaluating particular objects or events (e.g., college basketball players, the example in Appendix) to be represented by the degree to which any fuzzy variable (e.g., particular words for height such as tall or short, or words for intelligence such as brilliant or average) are relevant to that individual thing (e.g., desirability for the team). Rules can then be exercised based on combinations of multiple verbal phrases interconnected by logical AND and OR. Fuzzy logic contrasts with the more conventional

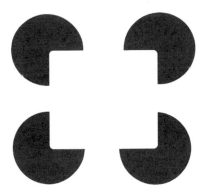

FIGURE 4.9 Kanizsa square illusion.

crisp logic where each object or event is represented in very specific objective terms (e.g., a basketball player having numerical height of 6 ft 6 in. and a grade point average of C). Evaluation by crisp logic is difficult because crisp logic rules are lacking for evaluation against many other individuals with differing combinations of the same attributes.

The advantage of fuzzy logic is that it is closer to the way real people think and do evaluations, whereas rules for comparing many items where each item is identified by specific objective identifiers are difficult for people to concoct. In his book, *Fuzzy Thinking*, Bart Kosco (1993) provides an excellent review of the ideas. He uses the Kanizsa square illusion (Figure 4.9) to suggest that certain unidentified aspects of a system are like the square that isn't there. Nice metaphor, but no scientific model there!

While the models described in this book are not explicitly couched in fuzzy logic, many of them could be, for reasons that have to do with introducing a large number of variables and human-judged approximations (and therefore relating to fuzzy logic). See Karwowski and Mital (1986) for a how fuzzy logic can be applied to human factors.

Fuzzy logic maps the meaning of words like tall and short, fast and slow, fat and skinny (and many other descriptors) into numbers that correspond to physical variables, depending on context. Those physical variables and the corresponding words can be used to characterize the observable objects or events of interest. However, many other words, such as love and hate and admiration and envy, indeed just about all words that express feelings, do not map upon physical variables very well, and there is no easy observable for the emotional words *per se*. Clearly, a facial expression in a photo or a drawing conveys an emotional state; one can relate emotional words to the curvature of the mouth, or eyebrows, and so on, but these I would categorize as metaphorical relations.

Fuzzy logic works with rules couched in ordinary words that are associated with physical observables, and derives a specific result based on such fuzzy rules; it does not work with metaphor. The mathematical example worked out in Appendix, Section "Mathematics of Fuzzy Logic" should make this clear.

SYMBOLIC STATEMENTS AND STATISTICAL INFERENCE (SEE ALSO APPENDIX, SECTION "MATHEMATICS OF STATISTICAL INFERENCE FROM EVIDENCE")

This section posits that any meaningful mathematic equations or statement in symbolic form is a model. Everyone is familiar with Newton's law, force = mass times acceleration, or $F = MA$, and most of physics is couched in mathematical equations. Even if a model takes its form in words, diagrams, or other graphics, certain of the properties can be coded by numbers, color, or any other symbols to distinguish attributes.

We call particular attention to modeling by statistics, and in particular statistical inference, since that is an accepted requirement of deriving a model from data in cognitive engineering experiments. Perhaps, the most widely accepted normative model for weighting evidence is the theorem nominally attributed to an English clergyman, Rev. Thomas Bayes (1763). It is called Bayes' theorem, and it results from simple algebra concerning probabilities of hypotheses, probabilities of associated evidence (data), and probabilities of contingencies between hypotheses and data. It is pure logic with the only assumption being that the hypotheses are mutually exclusive (only one is true) and collectively exhaustive (all possibilities are accounted for) and that the contingent probabilities are correct. The math for deriving and applying Bayes' theorem is given in Appendix, Section "Mathematics of Statistical Inference from Evidence."

The evidence from behavioral decision experiments is that people look for evidence to confirm already held subjective probability ratios (*confirmation bias*) and do not sufficiently weight available evidence according to the logic of Bayes' theorem. Bayesian models converge on one hypothesis or the other much faster than most people are willing to accept. In fact, many kinds of data disconfirming various stated beliefs and assertions are often available but are disregarded. Another pitfall in thinking revealed in Bayesian decision experiments is the tendency to use several very specific alternative hypotheses that are readily imagined plus a single "everything else" hypothesis, which does meet the requirement that the total set of hypotheses be mutually exclusive and collectively exhaustive. Often, however, insufficient consideration is given to those other hypotheses and kinds of evidence.

5

ACQUIRING INFORMATION

INFORMATION COMMUNICATION (SEE ALSO APPENDIX, SECTION "MATHEMATICS OF INFORMATION COMMUNICATION")

Information, when defined as the meaning of any given message, is essentially not possible to model, since it is totally dependent on context and the prior knowledge of the persons or systems sending or receiving the information. However, another definition of information, first developed by Claude Shannon (1949), and originally intended for use in telephone technology, is today broadly applicable as a scientific model.

The idea is simple, yet profound. The measure of information in a message is simply how unexpected the message content is, that is, the degree of uncertainty or possible variety (entropy) about the message as evident to the recipient before it is sent. After the message is received, if it is known with certainty, then the uncertainty or surprise is presumably reduced to zero. That difference in level of surprise (based on probability) determines the information transmitted from sender to receiver.

Appendix, Section "Mathematics of Information Communication" is the math to make the calculation of information transmitted as well as related

Modeling Human–System Interaction: Philosophical and Methodological Considerations, with Examples, First Edition. Thomas B. Sheridan.
© 2017 John Wiley & Sons, Inc. Published 2017 by John Wiley & Sons, Inc.

quantities called input information, output information, noise, and equi-vocation. The reader interested in applications of these concepts to human systems is referred to Sheridan and Ferrell (1974).

Human communications are coded in various forms, such as voice, gestures and "body language," handwritten and typed messages, pictures, mathematics, and all the methods described in Chapter 4, and transmitted directly or by artificial media such as the Internet, radio, or TV. The design of technology interfaces (displays and controls) is largely a matter of designing the form of communication that the message says what needs to be said to fit the needs of both the sender and the receiver. The receiver, of course, may be confused about how to interpret the meaning of the message, or else over-loaded with information. Display and interface design constitutes a very large part of cognitive engineering, but are beyond the scope of this book.

In ordinary human discourse, information communication is generally more complex than the transmission and reception of a single message. There are additional intervening steps that are typically carried out, as shown in Figure 5.1, representing the communication interaction between a sender and a receiver in a command operation. In practice, both parties play both roles, resulting in a sometimes complex and iterative interaction in an effort to make the most of the communication effort. Note the intermediate feedback shown by the lighter arrows.

There is another definition of information, often used as a measure of complexity of something. In essence, it is the length of a string of symbols of

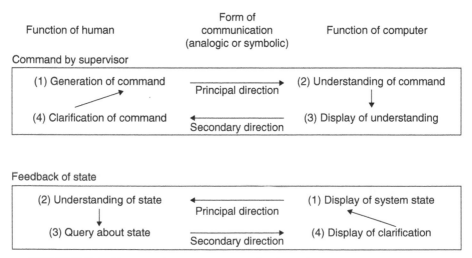

FIGURE 5.1 The complexity of communication with a person or a machine.

value 1 or 0 required to completely describe the thing. It is called the "Kolmogorov complexity," after the Russian mathematician.

INFORMATION VALUE (SEE ALSO APPENDIX, SECTION "MATHEMATICS OF INFORMATION VALUE")

Information value is based on a particular decision problem described by Raiffa and Schlaifer (1961) and later elaborated by Howard (1966) and others. It refers to a situation where the future is currently known only in terms of the probabilities of its possible states, but where research yet to be done would (ideally) reveal the future state with certainty. It also assumes that for any given future state the payoff for any action decision is known.

The question is, what is it worth to commit now (starting from only knowing the probabilities of alternative future states) to do the research necessary to know the exact future state (though right now not knowing what the research will reveal)? So information value is defined as the difference in what one can gain by the ability to know the exact future state (after incurring the cost of perfect research), compared to having to commit to action ahead of time knowing only the probabilities of alternative future states. Being ready to make spot investments or policy changes as events unfold is big business, and is also the basis for spending money on research. In other words, how much will research pay off? Without such readiness (or "insider information" on what is about to happen), a decision must be made that is not likely to be optimal as the future unfolds. An example with the mathematical details is provided in Appendix, Section "Mathematics of Information Value."

One application of this model is here and now: Reading this book takes time and effort, especially to overcome boredom that is understandably correlated with all this stuff about models. Is it worth your effort? Will whatever insights you would gain about future realities improve your ability to cope over and above not having those insights, and is the utility of that difference sufficient to overcome the time and effort costs of further reading? This of course is at issue for any educational effort. Under what circumstances does investment in education pay off? Presumably, even though at the time of investment the future remains unknown, the research eventually gives a clearer picture of the future, narrowing down the probabilities through better understanding. Education researchers are struggling to attach probabilities and values (utilities) based on observable evidence gleaned from case histories.

Information value can be computed in simple situations (Appendix, Section "Mathematics of Information Value"), but with imperfect research and complex multivariable problems if becomes essentially impossible.

LOGARITHMIC-LIKE PSYCHOPHYSICAL SCALES

An important property of the human nervous system is that information is coded as a logarithmic function of the physical property of the stimulus. More specifically, it is coded according to a metric on stimulus force or energy that is approximately a logarithmic function (or related inverse exponential function). The psychophysicist S.S. Stevens (1951) mapped such stimulus-to-sensation functions for loudness of sounds and brightness of lights, musical pitch, and similar functions for other senses. The term decibel, first named after Alexander Graham Bell, is also logarithmic, meaning it is a measure of sound intensity ratios. The difference in decibels between two sounds is defined by the log of the power levels such that

$$\text{The number of decibels}\,(\text{db difference}) = 10 \log\left(\frac{P_2}{P_1}\right),$$

where the log is base 10, the (original) "bel" difference being simply log (P_2/P_1), where capital P means sound power. The latter proved too big a unit, so the smaller decibel (1/10 bel) units became more popular (hence the "deci"). Since pressure magnitude is the square root of power, the number for decibel difference in terms of a ratio of sound pressure (p) magnitudes is twice as much, or $20\log (p_2/p_1)$.

An (approximate) logarithmic mapping is a very efficient mechanism for the nervous system, as it permits a very wide dynamic range of sensing. For example, the loudness of sound and brightness of lights span a range such that the greatest (on the verge of damaging to the organ) physical magnitude, compared to the smallest perceptible magnitude, is roughly 10^5 and equivalent for power is 10^{10}. This beats the dynamic range of just about any single artificial instrument, but with different instruments we can measure a much larger range. For example, on the electromagnetic spectrum, we routinely deal with a range between PET scans at 10^{21} Hz and AM radio waves at 10^5 Hz, a range of 10^{16}.

PERCEPTION PROCESS (SEE ALSO APPENDIX, SECTION "MATHEMATICS OF THE BRUNSWIK/KIRLIK PERCEPTION MODEL")

While correlation between two sets of items does not necessarily mean causality, it is strongly suggestive, and we tend to make the reasonable assumption of strong relationship. For a person to make judgments about some environmental state or property, there is a cascade of three transformations

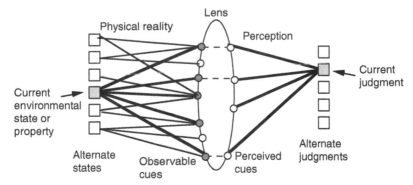

FIGURE 5.2 Interpretation of Brunswik lens model (after a diagram by Kirlik, 2006).

that must occur. First, there must be some physically observable cues or evidence that lends more credence to one hypothesis than another. Second, the human must access those cues (see, hear, or touch them) and weight their relevance to the discrimination to be made. Third, the human must be able to make use of the accessed and weighted cues to render a specific judgment. Figure 5.2 conveys the idea using the metaphor of an optical lens, the concept being a contribution of the psychologist Brunswik (1956) with quantitative refinement by Hursch et al. (1964).

The idea is that physical reality (at left) is seen through a "lens," and the effects (rays, heavy lines at left) of the present physical state or property of interest generate a set of cues (small dark circles on the left surface of the lens) that are potentially observable. At other times, other states generate cues also, some of them the same, some different from the present set (light circles on the left side). (This is the first transformation.) All cues from the present left-hand set are not necessarily perceived or properly weighted as relevant by the human (circles on the right surface of the lens), so that some cues are mistakenly perceived or weighted (dark circles not connected by dashed lines). In any case, the human ends up focusing on a set of perceived cues on the right side of the lens. (This is the second transformation.) The final transformation is using the perceived cues to make a judgment. (The correlation between states on the left and judgments on the right is the overall performance of the information acquisition plus judgment.) Kirlik (2006) has discussed this model with relevance to cognitive engineering.

ATTENTION

Wickens et al. (2003) published a model known as salience, effort, expectancy, value (SEEV) to go with his many experiments on attention. The variables that affect attention are all presented in Figure 5.3, with an attention-driving

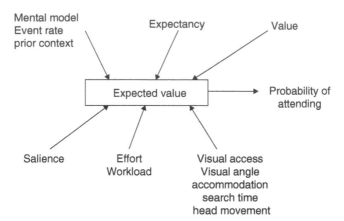

FIGURE 5.3 Wickens' SEEV model of attention. Source: Wickens et al. (2003). Adapted with permission of SAGE.

mental model, governed from the "top" by competing task demands (rate of change and prior signaling) as well as expectancy (EX) and value (*V*, importance), and from the "bottom" by factors of salience (*S*), effort (EF, workload), as well as head and visual movement constraints. Wickens offers a simple algebraic equation stating probability of attention (*A*) as

$$P(A) = K_s(S) - K_{ef}(EF) + K_{ex}(EX) + K_v(V),$$

where the *K* terms are coefficients.

The human factors and ergonomics literature has many examples of where lack of attention has led to accidents (see also Chapter 8, Section "Human Error"). For example, Moray et al. (2016) provide a detailed analysis of how field observations, experimental data, and modeling can be combined to explain how visual inattention to wayside signaling led to a major railway accident in the U.K. See also Moray and Inagaki (2014).

VISUAL SAMPLING (SEE ALSO APPENDIX, SECTION "MATHEMATICS OF HOW OFTEN TO SAMPLE")

Some kinds of attention demand are easily observable. When the major resource constraints are effort and decision-making in switching attention (including focusing and reading) from one relatively simple display to another, simplest assumptions are that observation time is constant across displays and that the time to transition from one display to another is negligible. In this special case, the strategy for attending becomes one of relative sampling frequency. That means checking each of many displays often enough to ensure that no new demand is missed, and/or sampling a continuously changing signal often enough to be able to reproduce the highest frequency component present, the bandwidth.

Where the bandwidth (highest frequency component) of any display input signal i is ω_i, the required sampling frequency, called the Nyquist frequency, is $2\omega_i$. That can easily be shown by considering that a Fourier series to represent any function of time duration T has two coefficients for every frequency multiple of the fundamental frequency, $1/T$, up to the bandwidth ω_i. This is $2\omega_i T$ coefficients per T, or a time-average $2\omega_i$.

A classical experiment by Senders et al. (1964) made use of the aforementioned idea. He showed that experienced visual observers do tend to fixate on each of several displays in direct proportion to each display's bandwidth. Figure 5.4 shows some of the results. Each of five highly practiced subjects monitored four zero-centered instruments, each having signals of different bandwidth ω_i. Signals were actually sums of noncoherent sinusoids, a technique that makes them appear random. The instruments were located at the corners of a square, and were separated by approximately 60° of visual angle. The total Nyquist number of fixations, $\Sigma(2\omega_i)$, was selected so as not to overload the subject. The task was to report when any signal reached extreme values.

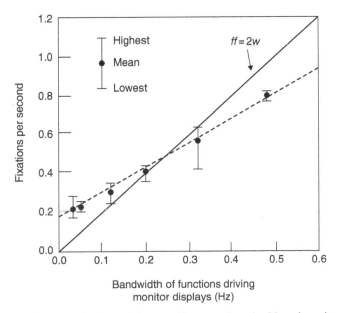

FIGURE 5.4 Senders' model: sampling matches the Nyquist criterion.

The model predicts that the data should lie on the solid diagonal line (i.e., $2\omega_i$), if a sample is instantaneous. Note that subjects tended to over-sample the lowest frequency displays and undersample the highest frequency displays—a "hedging toward the mean" that is found in other forms of decision behavior. The undersampling at higher frequencies could have been because samples were not instantaneous (actually they were relatively constant at about 0.4 s, independent of frequency), and therefore the subjects were getting some velocity information too. If on each sample they got perfect velocity information, they needed to sample only at ω_i. Clearly, they compromised between the ω_i and $2\omega_i$ rates; the slope for Senders' data is 1.34. The oversampling at low frequencies (long time delays between samples) might result from forgetting, pointing to other evidence that very-low-frequency signals can be sampled at rates much in excess of $2\omega_i$.

Predicting sampling on the basis of the Nyquist interval makes several assumptions: (1) that averages are over a long time period (nominally infinity); (2) that all displays are of equal importance; (3) that there are no costs for sampling; and (4) that the observer has perfect knowledge of the bandwidths.

In many tasks, the usefulness of information is a function of how recently a particular display or source of information has been observed. At the instant of observing a display, the operator has the most current information, but as time passes after one observation (action) and before a new one the information becomes "stale," eventually converging to some statistical expectation. How often should a person sample to gain information and readjust controls, given that there is some finite cost for sampling as well as the gradually increasing cost of inattention? A more elaborate model for this situation, together with the mathematics, is described in Appendix, Section "Mathematics of How Often to Sample."

SIGNAL DETECTION (SEE ALSO APPENDIX, SECTION "MATHEMATICS OF SIGNAL DETECTION")

The signal detection model has seen wide application within cognitive engineering (Green and Swets, 1966). It characterizes how people or electronic systems succeed or fail in detecting signals when the signals are corrupted by noise or are otherwise faint (near the threshold of seeing, hearing, or touching). The "noise" can be visual, auditory, or tactile, or can be some form of mental distraction. In computer or other complex systems, noise can take various other forms of signal corruption. While this model was originally developed for radar applications, it has been a key model for analyzing and reporting experiments in human psychophysics of sensing and perception.

The model includes prior statistical knowledge of the (overlapping) probabilities of signal and noise relative to the strength of some evidence. It also includes the payoff matrix (rewards for true positive response and true negative response and penalties for false positive and false negative). Finally, it includes a factor that accounts for whether one is conservative (avoiding risk) or liberal (accepting risk).

One interesting result of many experiments is that people change their tendency to make detections (indicate what they "truly" see or hear) according to how much they are rewarded for each choice, even for very basic judgments like hearing a tone or seeing a light near the threshold of hearing or seeing. The mathematics, derived from Bayesian logic, to specify the decision criterion (whether to decide signal or noise) as a function of probabilities, rewards and costs, is given in Appendix, Section "Mathematics of Signal Detection."

SITUATION AWARENESS

Situation awareness (commonly referred to as SA) is a construct in cognitive science that is widely discussed but very hard to measure. SA is defined by Endsley (1995a,b) as "the perception of environmental elements with respect to time or space, the comprehension of their meaning, and the projection of their status after some variable has changed, such as time, or some other variable, such as a predetermined event." More simply stated, it is "knowing what is going on around you." Endsley and her colleagues have performed many studies of SA, mostly in military and commercial aviation contexts and highway driving.

How to measure SA? Endsley developed a technique called Situation Awareness Global Assessment Technique (SAGAT). In a simulation, the experimenter abruptly and unexpectedly stops the simulation and asks the subject questions about what the subject saw or thought, which indirectly allows inferences about what was known or unknown about the situation— how "aware" was the subject. This method is widely used, but it leaves a wide margin for effects of experimental design and experimenter behavior to affect the result, not dissimilar to difficulties in efforts to infer mental models to be discussed in Chapter 10.

Wickens et al. (2012) combine notions of attention with those of SA in modeling pilot error. They relate SA to their SEEV model of selective attention, mentioned in this chapter, Section "Attention". Indeed, there can be no SA without attention.

MENTAL WORKLOAD (SEE ALSO APPENDIX, SECTION "RESEARCH QUESTIONS CONCERNING MENTAL WORKLOAD")

The origins of the concept of mental workload go back to the early time-and-motion studies in factories, and to the "scientific management" of F.L. Taylor (1947). Even 100 years ago, Yerkes and Dodson (1908) modeled how arousal (an alternate term for mental workload) affects performance by means of an inverted U curve, now well-known as the Yerkes–Dodson law. Too little task load makes a person bored and inattentive, and too much makes him overloaded. Mental workload has been an especially popular topic in human–machine systems research for the past four decades (Moray, 1979). Much of this research has been focused on pilots, initially to resolve disputes about whether the commercial aircraft flight deck should have a crew of two or three. Motivation was also to decide on the limits of what can be assigned to fighter and helicopter pilots flying nap-of-the-earth combat missions. It has also been of concern in nuclear power plant control rooms and other settings where overload, boredom, or other concomitant forms of deterioration of operator performance pose a serious threat.

It is not mental workload *per se* that is of ultimate concern. It is whether the human (e.g., pilot or other operator) can perform the given task with sufficient margin of attention to compensate for unforeseen circumstances, stress, or complexities which may be infrequent but certainly do arise from time to time. Thus, operator performance, which translates to system performance and safety, is the ultimate concern. As "objective" workload (taskload) factors mount (i.e., variables such as number of tasks, complexity of tasks, urgency, and cost of not doing the task properly and on time), mental workload (whatever it is) surely increases. But, unfortunately, until these objective loads reach a high level, operator performance (as measured by system performance) may show little, if any, decrement. Then, with a slight further increase, performance may decrease precipitously. At that point, subjective mental workload is particularly hard to measure, but may also decrease. The hope has been that physiological and even subjective indices of mental workload can provide a better predictor of performance breakdown, well before the latter occurs, than direct measures of performance. The upper plot in Figure 5.5 illustrates the point.

Some definitions and discriminations such as the following are relevant:

Physical workload is the energy expended (e.g., in calories) by the operator in doing the assigned task, measured by the body heat given off or by the conversion of respired oxygen to carbon dioxide. Alternatively, it can be the net mechanical work done on the environment (e.g., lifting a

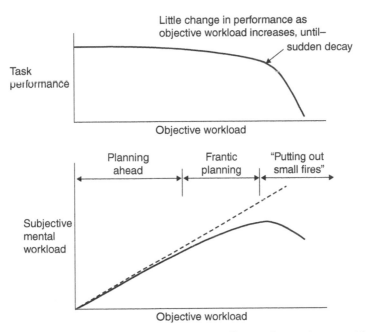

FIGURE 5.5 Properties of mental workload (effects of very low workload not shown).

10-pound weight 5 ft produces 50 ft-pounds of work). But physical workload is essentially different from mental workload, sometimes contributing to it (heavy physical demands can produce mental stress) and sometimes reducing it (exercise can be relaxing).

Task performance, as explained earlier, is what we seek to predict by measuring and modeling mental workload. However, it is best not to regard task performance as a legitimate measure of mental workload. An experienced operator is likely to be able to perform a task well with little mental workload, while an inexperienced operator may perform poorly with great mental load. Therefore, task performance and mental workload should be kept distinct from one another.

Task complexity is a property of the task independent of the human operator or the operator's behavior. Examples of task complexity are number or variety of stimuli (displays, messages) or actions required, rate of presentation of different displays or required pace of actions, time urgency before a deadline, improbability of the information presented or the actions required, and the degree of overlap of multitask memory demands. The latter notion is illustrated in the lower diagram of Figure 5.5. Measures of task complexity are properties of the task, not of the person. Hence, they can be called measures of "objective workload," "imposed workload," or "task load," but not of "mental" workload. The following three measures can be associated with the operator's mental workload.

Subjective rating is by definition a nonobjective measure, but interestingly it is the standard against which all objective measures are compared. Subjective mental workload rating can be done according to a single-dimensional or a multidimensional scale. A single-dimensional scale in a form similar to the test pilot's Cooper–Harper handling quality scale (Cooper, 1957) is commonly accepted and has been widely used for pilot workload.

Sheridan and Simpson (1979), after systematically observing pilot behavior riding "jump seat" on a number of airline flights, proposed a multi-dimensional scale using separate dimensions of time pressure ("busyness"), mental effort (problem complexity), and emotional stress. Such a scale was refined by Reid et al. (1981), used by the Air Force, and is called the *subjective workload assessment technique* (SWAT) (O'Donnell and Eggemeier, 1986). About the same time, NASA developed a similar scale called TLX.

It is important to consider that an operator's rating of mental workload will not necessarily be the same as the experience of mental workload. There are factors that drive the rating to differ from the conscious experience, and there are aspects of the experience that are not conscious. Automation, if it has no other effect, certainly changes the nature of the experience.

Physiological indices used for mental workload assessment are many. Some of these are heart rate, heart rate variability, changes in the electrical resistance of the skin due to incipient sweating, pupil diameter, formant changes of the voice (especially increases in mean frequency), and changes in breathing pattern. Great variability has been found in these measures; no single measure has been accepted as a standard.

Secondary task performance measurement requires that the experimental subject do a secondary task at the same time the subject is performing the primary task. Examples of secondary tasks are backward counting by threes or sevens, generation of random numbers, and simple target tracking. The better is performance on the secondary task, so the argument goes, the less the workload of the primary task. The problem has been that airplane pilots and operators performing actual critical tasks sometimes refuse to cooperate in such measurement, although in simulators (where they know there are no real dangers) they may be cooperative.

Aircraft piloting, nuclear power plant operation, and military command and control can involve long periods of boredom punctuated by sudden bursts of activity, without much demand for activity at the much-preferred intermediate levels. To some extent, this is true even if no automation is present (e.g., long periods of manual piloting to maintain course), but automation only exacerbates the problem. It is not that the operator is simply bored doing a manual task. She may not be involved at all for long periods, and may become "decoupled" even from feeling responsible for control, knowing a computer is doing the job. But then, when problems arise that the automation cannot handle, or the automation breaks down, the demands on the operator may suddenly become severe.

Such transients, "coming into the game cold from the bench," so to speak, pose serious risks of human error and misjudgment. Experienced operators will react instinctively and usually be correct, or if the situation is novel may try to "buy time" to collect their wits. Operators have been known to accommodate to information overload by bypassing quantitative input channels (e.g., instruments) and going instead to *ad hoc* qualitative channels (e.g., voice messages), which may not be reliable. Further discussion of mental workload is found in Appendix, Section "Research Questions Concerning Mental Workload."

At very low mental workload, what might be called *hypostress*, human performance is known to decrease since the human becomes bored, distracted, and/or falls asleep. Such low levels of cognitive demand are not depicted in Figure 5.5, but are shown in Figure 5.6, based on the work of Hancock and Szalma (2008), an extension of the well-known Yerkes–Dodson law mentioned before. Too little or too much mental load (mental stress, arousal) leads to poor performance, but for different reasons in each case. In between is a "comfort zone" or "sweet spot." Those authors discuss a somewhat wider zone of *psychological adaptability* (shown on the vertical axis) and an even wider zone of *physiological adaptability*, an absolute limit, outside of which performance drops rapidly to zero.

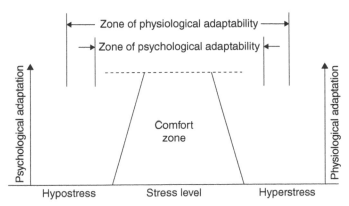

FIGURE 5.6 Regions of workload accommodation. Source: Hancock and Szalma (2008). Adapted and modified from Elsevier.

EXPERIENCING WHAT IS VIRTUAL: NEW DEMANDS FOR HUMAN–SYSTEM MODELING (SEE ALSO APPENDIX, SECTION "BEHAVIOR RESEARCH ISSUES IN VIRTUAL REALITY")

In recent years, new kinds of human–system interaction have begun to demand attention by modelers in the cognitive engineering world, namely the experience of virtual reality (VR) and augmented reality (AR). VR refers to humans interacting with a computer-generated (and therefore virtual) environment. Typically, the observation is through either a head-mounted or an environment-mounted visual display. The experience can be so compelling that the observer loses track of the real surrounding environment. AR simultaneously mixes virtual images with real scenes (again through computer processing)—adding features that are not actually in the real environment.

Since new helmet or eyeglass displays have sensors that detect which way the head is pointing (up or down, left or right) the computer can sense this and make the computer-generated image be what the viewer would see if the head were orientated in the same way while viewing the real environment. This is very compelling and produces the striking sensation that VR researchers call "immersion" in the virtual environment.

Let us admit that the term "virtual reality" is an oxymoron, a self-contradiction from the normal use of English words. If something is virtual, it is not real, and if it is real, it is not virtual. But the term has caught on, and is accepted in the current vernacular, so we will use it.

Both VR and AR enable a new kind of human–system interaction and potentially many new opportunities for display of information from real operating systems or from models. One of the earliest uses of VR was in flight simulators that presented not only a compelling visual scene corresponding to the result of pilot actions, but also included six-axis motion simulation. Another early use was by architects, to allow people wearing head-mounted displays to "walk though" virtual buildings, look around 360°, and experience them before they are even built. An early use of AR was to allow mechanics to examine actual machinery and see labels or instructions visually "attached" to the components they are looking at. One AR proposal is to allow driver trainees to experience suddenly appearing (virtual) obstacles that are potential collisions while actually driving on an otherwise normal roadway. Figure 5.7 shows two AR images from a video of a virtual truck emerging from a side street, where all the rest of the scene is real background (from an actual road test of the idea by Sheridan, 2007).

FIGURE 5.7 Two images of a video showing superposition of computerized truck images on actual driver view in a test drive on a country road. White objects on trees along the roadway are fiduciary markers to enable continuous geometric correspondence of the AR image to the real world.

VR technology has motivated behavior research into how humans perceive "where" they are present and what they perceive as reality. The research helps us understand how easily we can be made to experience what is not really there.

There are several "quality" measures that can be applied to virtual environments (Sheridan, 1992b, 1996). One type of measure is the sense of reality—how "immersed" does the human subject feel, measured on a subjective scale—which is important in entertainment applications. However, it may or may not be important in more practical applications (e.g., flying the flight simulator, performing virtual surgery, checking to make sure the components fit together in a virtual instantiation of a mechanical design or building). In these practical applications, one is more concerned with operator learning, so both performance in the virtual environment and transfer of that learning to the real environment are what is important.

With respect to the experienced sense of visual reality, there are three factors that contribute to "immersion":

1. The *display information rate*, consisting in turn of
   a. spatial resolution (defined in *pixels* for vision, *taxels* for touch);
   b. temporal resolution (frame refresh rate in images per second); and
   c. magnitude resolution (binary bits per pixel embodied in color or gray scale).

   Note that the mathematical product of these three quantities (pixels per image, images per second, and bits per pixel are equal to bits per second, the key metric for rate of information transfer).

2. The *ability to move about* and orient in the sensory space (e.g., to scan a visual display with the eyes, to actively move the hand across a tactile display, to move the body within a virtual room).
3. The *ability of the subject to effect changes* in the virtual environment (e.g., to manipulate objects in the virtual world).

If what is seen, heard, or touched does not correspond realistically to scanning or viewpoint or bodily changes, the sense of presence in the virtual environment is lost. This loss can occur if there is sufficient delay between movement and display response (more than 0.1 s), which can occur if too much computing is necessary to effect timely and accurate movement in the displayed transformation.

We have ample evidence that the sense of presence ("immersion") in the virtual environment is enhanced when multiple senses are employed simultaneously (e.g., vision plus hearing, vision plus touch, or all three sensory modes). This is in contrast to the subject perceiving simply that he/she is physically located in a laboratory and simply being asked to play an artificial game. Figure 5.8 (top) conveys this inter-relationship. Figure 5.8 (bottom) suggests relevant independent and dependent variables.

Auditory VR technology is such an old story that we do not recognize it as VR. Apart from story-telling on the radio, we have electronically recorded music, enhanced to produce a stereo effect by using earphones or strategically placed loudspeakers corresponding to placement of microphones at the recording site. More recently, a scientific understanding of how the shape of the ear enables us to distinguish sounds behind or above the head has resulted in added electronics to replicate this effect. For example, a listener can be made to experience a virtual buzzing insect swirling around the head at easily identifiable locations, where the auditory "image" of the insect is produced electronically as sound characteristics in the earphones.

New technology also permits haptic immersion in a virtual environment, where the participant can experience the sense of touching objects in the environment that are not really there. How can this possibly be? It is achieved by having the participant wear a specially made glove that stimulates the tactile nerve endings on the skin in correspondence to whether the computer software indicates contact with an object at any particular location of the actual (location measured) hand.

Using similar technology, the participant's hand can be placed within a mechanism that causes an actual remote robotic hand/arm arbitrarily far away to touch or handle real objects. The human hand can receive force feedback corresponding precisely to the forces the remote hand encounters in its

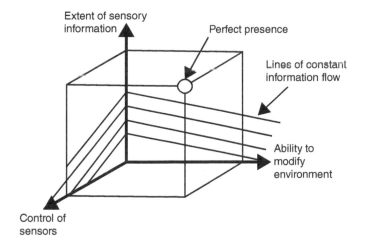

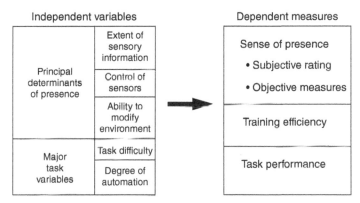

FIGURE 5.8 Variables contributing to "presence" in VR.

(remote) environment. This technology for remote handling was originally applied to enable a human operator to peer through a leaded glass (shielded) window to manipulate objects in a nearby radioactive "hot" environment. Without the virtual touch and force feedback, it would be impossible to perform such telemanipulation safely and reliably.

It is interesting to consider the close relationship between VR resulting from clever computer graphics and human operation of an actual remote robotic hand or vehicle, where an important goal is to achieve a sense of *telepresence* (feeling of being present at the actual remote location). In both cases, the quality of the interface is responsible for generating a mental model that engenders *telepresence*. Figure 5.9 conveys the idea.

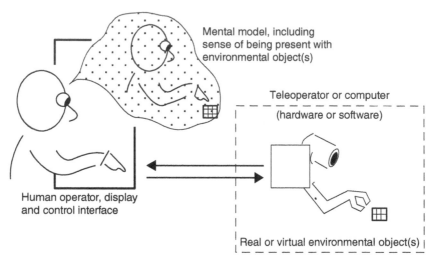

FIGURE 5.9 Relationship of VR created by computer and telepresence resulting from high-quality sensing and display of events at an actual remote location. The dashed line around the remote manipulator arm suggests that the remote arm can be either real or virtual, and that if the visual and/or tactile feedback are good enough, there will be no difference in the human operator's perception (mental model, shown in the cloud) of the (real or virtual) reality.

6

ANALYZING THE INFORMATION

TASK ANALYSIS

A standard human factors technique is task analysis. A task analysis can be said to be a model of the task.

In addition to the form such as the one shown in Figure 6.1, task analyses are often represented as flow charts that can show contingencies (for multiple paths that are dependent on fulfillment of given conditions), feedback loops, and so on. Common instructions on how to assemble products (delivered in parts) often complement the task analysis using simple pictures with arrows. Computer programs are a form of task analysis for the computer. Sometimes, a task analysis of what a human operator is required to do is interlinked on a step-by-step basis with a description of what a computer is supposed to do. Detailed task analyses of this type are developed as a common communication basis for engineers, trainers, and pilots regarding operations of aircraft and spacecraft (Kirwan and Ainsworth, 1992).

Modeling Human–System Interaction: Philosophical and Methodological Considerations, with Examples,
First Edition. Thomas B. Sheridan.
© 2017 John Wiley & Sons, Inc. Published 2017 by John Wiley & Sons, Inc.

Task step	Operator or machine identification	Information required	Decision(s) to be made	Control action required	Criterion of satisfactory completion

FIGURE 6.1 A hypothetical form for performing a task analysis.

JUDGMENT CALIBRATION

Expert and/or user judgments are critical in designing a machine or a policy. One of the most straightforward problems of extracting judgment information from a person is the degree to which the judgments people render about the values of variables in the physical world are biased. This problem has a solution, for those judgments can be made more accurate and therefore useful by the process of *calibration* (Mendel and Sheridan, 1989).

There are several key assumptions about calibration: (1) Mental models and processes are relatively stable statistically, so that individual judgment biases about the same or very similar variables in the same or similar situations can be treated in the same way, that is, as a probability density. (2) The human judge can not only render her "best guess" (mean of the judgment density function for that event) but can also provide fractiles (percentiles) (say 90% confidence limits) on the judgment, as a way of specifying "how sure she is." (Machines, by the way, are generally not capable of this, unless of course they possess the rich store of contextual data from which humans make judgments.) (3) Eventually, the judgments can be verified against events in the physical world (eventually the truth is known).

Assume that a number of judgments are made that relate to the same "situation." They can be judgments of the same variable at different times or places, or judgments of different variables, but all involve the same mental model. In this case, except for scaling factors, the biases and precision of judgments on the different variables are likely to be the same. In each case, the judge is asked to provide not only the best estimate (median) but also the 90% confidence limits. The first three such judgments are represented by the vertical ticks on the top three lines in Figure 6.2 (scales for arbitrary variables

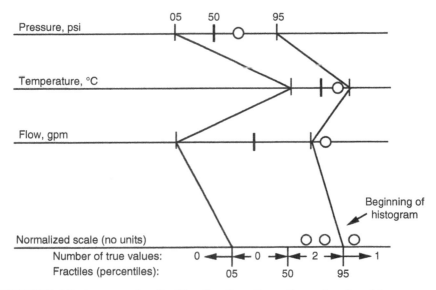

FIGURE 6.2 An example of calibration for a three-dimensional problem space.

chosen for this example). Heavy ticks are medians and light ticks are 90% confidence limits. Eventually, the truth about each judgment (encounter with real event) is determined, and that is recorded for each case (as the circle). After a number of such comparisons, the results are aggregated on a common normalized scale, where the 90% confidence ticks are the same, and the units of each variable lose their meaning (see bottom scale in Figure 6.2). Now the true values form a frequency distribution, and statistical estimates can be made of the median and the standard deviation.

If the judge had no biases, either in her "best estimate" or in her 90% confidence estimates, the median of the true values would lie at the 50% judgment fractile tick. The offset of the 50% judgment fractile relative to the median of the true values is the expected bias. If the 90% fractiles of the true value histogram lie outside the 90% subjective judgment ticks, that implies that the judge has more confidence than is warranted. If the 90% fractiles of the true value histogram lie inside the 90% subjective judgment ticks, that implies that the judge has less confidence than she should have in her own judgments. A chi-square statistic adapts easily to testing whether the expected number of true values appears in each of the interfractile intervals. Mendel and Sheridan (1989) extend these ideas for getting calibrated "best measures" from multiple human experts or other information sources.

VALUATION/UTILITY (SEE ALSO APPENDIX, SECTION "MATHEMATICS OF HUMAN JUDGMENT OF UTILITY")

The common meaning of the term *utility* is "usefulness" in a very general sense. But in economics, it means relative subjective worth. Relative worth is different from money. If you are a rich person, the utility of the next dollar is less than if you are a starving person, though the dollars are physically the same in either case. If you have just eaten a large bowl of ice cream, the utility of another bowl is less than the utility of the first bowl, even though the dollar cost of the next bowl is the same as that of the first.

How to define and measure utility? This problem of relative worth and fairness in economic welfare intrigued philosophers such as Jeremy Bentham and John Stuart Mill for 100 years. However, Von Neumann and Morganstern (1944) defined utility by an axiom and an experimental procedure, and this definition is now universally accepted.

The procedure is straightforward and requires the experimental subject (whose utility or relative worth is being determined) to make comparative relative worth judgments between three things (objects or events). One thing is initially assumed to have high value, one has a low value, and the third has a value somewhere in between. The procedure results in locating where the value of that third thing lies on the utility scale. From repeated trials, one can derive a utility function for a large set of things, starting from the given high and low limits, and mapping the relative worth against the physical variable, as explained in Appendix, Section "Mathematics of Human Judgment of Utility."

In economics and engineering, one is often seeking *multiattribute utility functions*, mathematical equations that define relative goodness in terms of different weightings on a number of salient attributes of the thing being chosen. For example, the goodness of a teacher might have to do with intelligence, preparation, and personality, to name only a few attributes. Cars have attributes of passenger capacity, fuel economy, comfort, and style. For any choice, one can keep thinking of additional attributes that might be relevant. Keeping the attributes to a tolerable number is essential.

If each attribute X were independent of the others and were scaled 0–1 from worst to best, and a judge could decide on relative weights C for how important that attribute is (to that judge), then a simple-minded valuation would be

$$U = C_1 X_1 + C_2 X_2 + C_3 X_3 + \cdots$$

However, there are numerous methodological issues with such an approach, such as what is included in the set of attributes, how they are defined, are the attributes really independent, how the individual attributes are scaled within the range 0–1, and, more difficult, is goodness really simply a linear sum of weighted attributes or some more complex function. The interested reader is referred to Appendix, Section "Mathematics of Human Judgment of Utility" for a treatment of single attribute utility from a mathematical perspective, or to Sheridan and Ferrell (1974) for derivation of mathematical expressions for multiattribute utility.

Utility models are invaluable in deciding on how different people, representing different demographic backgrounds (interests, educational, cultural, or economic), assign *different* relative worth to different beliefs, public policies, and so on. Such models are widely used in management, marketing, policy planning, and system engineering. In control engineering, another term for utility function is *objective function*, the function to be maximized in doing control.

Earlier, we had noted that human perceptual magnitude of basic physical variables follows a logarithmic-like function relative to physical magnitude (usually called the Weber–Fechner function after its early discoverers). It is more or less true that subjective worth or utility of wealth follows a similar function, that is, satisfaction tapers off dramatically with greater wealth. This suggests that equalizing wealth across a population would lead to greater average utility.

RISK AND RESILIENCE

Definition of Risk

Risk of some event is normally defined to be the probability of failure multiplied by the cost of the consequences of that failure. To have any meaning, both terms must be defined in terms of the context of the situation. If there are multiple failure modes, risk can be considered to be the sum over all failure modes of probabilities times costs.

Meaning of Resilience

In recent years, there has been a trend by some researchers of risk and safety in systems involving humans to move away from traditional risk quantification. The move is toward what is called *resilience engineering* (Woods et al., 2012). It presupposes that human errors and machine failures will probably occur, and a more useful effort should be placed on means to anticipate disturbances and be able to recover and restore the system to the original state or, if need be, some acceptable state that is different but still safe.

Resilience engineering is really a family of ideas. It has been defined variously as follows:

- Worker, team, and/or organization ability for early detection of drift toward unsafe operating boundaries
- Worker, team, and/or organization flexibility and adaptability in responding to surprises so as to mitigate any undesirable consequences
- A system's ability to anticipate and respond to anomalous circumstances so as to maintain safe function, recover, and return to stable equilibrium (to the original operating state or to a different state)
- Motivation to continue imagining and worrying about "what's happening" safety-wise when "nothing (apparent) is happening" and the system seems safe

One popular metaphor of resilience is the common stress–strain curve of solid mechanics (see Figure 6.3). Many physical materials, when stretched by an external force (strained), from some length A to a new length B, develop an intermolecular stress (force) that enables the material to return to its original length when the external force is removed (in other words, a perfect spring). But when stretched beyond an elastic limit to C, the intermolecular stress will "give" and not increase linearly with stretch. In that case, when the external force is released, the material will not return to the original length but return to a different point D. For mechanical engineers, it is critical to

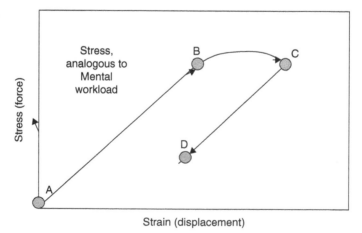

FIGURE 6.3 Stress–strain analogy to resilience.

know what the elastic limit is in terms of applied force and stretch, and for the safety engineer, it is critical to know the elastic limit in terms of task load and accompanying stress and mental workload.

TRUST

One can cite multiple factors that engender trust, whether in people or in automation. Most obvious is reliability, whether a person or system has consistently performed well or as expected, and did not fail. Immediately following a failure, we are inclined not to trust. This can be irrational in the case where otherwise reliable systems have rare chance of failures and there is evidence that the person or institution takes pains to acknowledge the failure and makes serious efforts to prevent such failures in the future. If the failure was in the long past and recent performance has been successful, we are inclined to trust.

We tend to trust people and things that are familiar, events that we think we understand, and have experienced. For this reason, we adhere to beliefs and cultural norms held by our parents and friends growing up—in short, our tribe. Unfamiliar people, systems, and situations provide insufficient statistical evidence on which to base trust. We also tend to trust people and systems that we depend on, perhaps because we must (we may have no choice), but also because those entities are familiar. Commercial advertising is a greater determiner of what we trust than we are willing to admit. Trust may be based on evidence that is not direct but is second hand. Having initial trust in a person or institution, we are inclined to trust something that person or institution says.

Experiments and models of trust are now becoming more popular in various fields of science and technology, especially in psychology (interpersonal trust), in computer science (cyber-security, and operation of complex systems such as air traffic control), and in political science.

Hoffman et al. point out (2009) that trust has aspects of: (1) an *attitude* (of the trustor about the trustee); (2) an *attribution* (that the trustee is trustworthy); (3) an *expectation* (about the trustee's future behavior); (4) a *feeling* or *belief* (faith in the trustee, or a feeling that the trustee is benevolent, or a feeling that the trustee is directable); (5) an *intention* (of the trustee to act in the trustor's interests); and (6) a *trait* (some people are trusting and more able to trust appropriately).

Lee and See (2004) provide an extensive review of the trust literature as well as a framework to represent the many relevant factors of predictability, dependability, reliance, performance, context, emotion, analysis, information

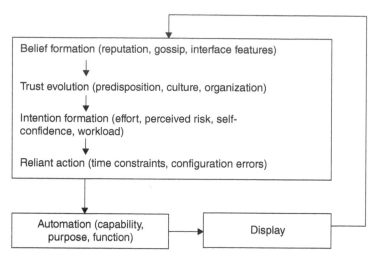

FIGURE 6.4 Variables affecting trust (after Lee and See, 2004).

accessibility (display), and so on. A rational person will calibrate level of trust (assuming there is opportunity for sufficient data gathering) so as not to overtrust or undertrust. Greater trust is typically correlated to more extensive capability of the automation. The relationship of the many factors is presented in their model as a dynamic closed loop. Figure 6.4 suggests the interplay of variables affecting trust. More recently, Schaefer et al. (2016) provide a meta-analysis of factors influencing the development of trust in automation.

7

DECIDING ON ACTION

WHAT IS ACHIEVABLE

It may be useful to render a problem space in some form of graphical representation that allows the decision maker to visualize in multiple dimensions
(of the salient variables of the problem) what is *achievable* within the
problem space of what is *ideal* and what is *acceptable*. Some solutions may
be achievable, given available resources (time, energy, money, etc.) but not
be desirable, while some may be acceptable but not achievable. If there were
only two dimensions to the problem space, this would create a rectangle (as
shown in Figure 7.1). The larger rectangle is defined by what is desirable
and what is acceptable, based on inputs by the human decision maker. Points
within this space can be evaluated as to achievability by a computer armed
with relevant physical laws, economic, or legal constraints that simultaneously
determine the smaller rectangle.

In general, problems (designs, policies) are constituted by a much larger
set of variables N, and so these nested spaces must be represented in N dimensions, impossible to visualize in a flat rendering. A computer, of course, has
no practical limitation on the size of the problem space, so any graphical
representation may have to show the ideal, acceptable, and achievable points

Modeling Human–System Interaction: Philosophical and Methodological Considerations, with Examples,
First Edition. Thomas B. Sheridan.
© 2017 John Wiley & Sons, Inc. Published 2017 by John Wiley & Sons, Inc.

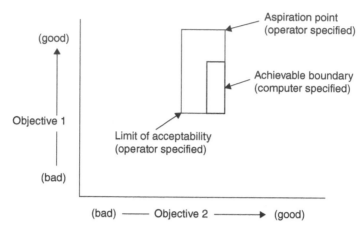

FIGURE 7.1 Example of determining the space of what is achievable within the space defined by what is aspired to and what is acceptable (in a simple two-dimensional problem space).

for each variable as marks on a line for each variable. This computer technique (Buharali and Sheridan, 1982) provides a model for aiding decision visualization to assist the decision maker or designer in this regard. There are many computer processing techniques for multivariable analysis that embody constraint equations, such as linear programming.

DECISION UNDER CONDITION OF CERTAINTY (SEE ALSO APPENDIX, SECTION "MATHEMATICS OF DECISIONS UNDER CERTAINTY")

Now we look at models of how people make optimal decisions. Decision under certainty means everything is known and available; there are no probabilities to contend with, only the selection among available alternatives. Which cereal on the supermarket shelf to buy? Which car in the showroom to purchase? Which girlfriend to propose to?

The answer presumably is to select the alternative with the greatest utility. But, as indicated in Chapter 6, Section "Risk and Resilience," this demands selection among alternatives that differ in many ways—a problem of multi-attribute utility, as mentioned earlier. So the need is to envision the alternatives in juxtaposition with respect to their attributes and relative utilities. This will help determine which alternatives can be cast aside and which come close to having the same utility. Sometimes, gut feelings come into play that are not accounted for in the initial utility judgments, and may dictate modifications in the utility picture. The considerations in making such decisions are best described by means of a quantitative example, as in Appendix, Section "Mathematics of Decisions Under Certainty."

DECISION UNDER CONDITION OF UNCERTAINTY (SEE ALSO APPENDIX, SECTION "MATHEMATICS OF DECISIONS UNDER UNCERTAINTY")

When the states of the world and consequences of each decision are known only by probabilities that are less than one (the decision must be made without knowing for sure what outcome will happen), then we are talking about decision under uncertainty. Under uncertainty, any decision option has attached to it alternative consequences, each of which has a probability.

Here the decision maker is stuck with having to weigh the relative positive or negative value (utility) of the consequences of each decision alternative if those consequences occur, along with the probabilities that they will occur. How to do this? One can decide on whether to (1) be very conservative and look only at the worst that can happen, or (2) to consider the consequences, good or bad, in direct proportion to their probabilities, or (3) something in between.

Again, these relationships can best be seen through an example, see Appendix "Mathematics of Decisions Under Uncertainty."

COMPETITIVE DECISIONS: GAME MODELS (SEE ALSO APPENDIX "MATHEMATICS OF GAME MODELS")

Game models represent decision-making in a competitive environment—sports, business, war, or personal competition. They model a situation where each competitor (normally two) must make a decision not knowing what the other competitor will decide, but where the outcome for each (same or different) depends on the combined decisions of both (or all) parties.

If both parties decide on war they will see battle. If one decides on war and the other decides on peace, the former will conquer the latter. If they both decide on peace, both sides will benefit, but not as much as a winner getting the spoils of war. What strategy should each play?

Both sides in any game situation have their ideals, and act according to what they perceive will net them the greatest payoff. But they may not take into account how the payoffs change after repeated plays—due to changes in opinion (e.g., development of trust or distrust) by the players or forces from the outside. Evidence shows that moves made in repeated plays normally do depend on the outcome of immediate past plays, typically with a time delay. The trust or lack thereof that develops is therefore dynamic. However, sometimes there is no second chance.

On a happier front, one can model how consistent acts of kindness by one party can over time convert some other party (say an urban gang member) from taking every selfish advantage to reciprocating with love and cooperation.

Game theory (Von Neumann and Morganstern, 1944) is best illustrated by example. Some classes of formal games are illustrated in Appendix, Section "Mathematics of Game Models."

ORDER OF SUBTASK EXECUTION

Tulga and Sheridan (1980) developed a model of attention allocation among multiple randomly appearing task demands, based on an abstract graphic presentation. The "model" here is really the diagram of the experiment, which was also the display the subject saw.

Tasks were displayed to the subject on a computer screen as is represented in Figure 7.2. New tasks appeared at random times and locations on the

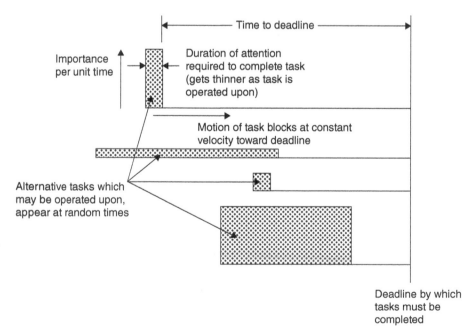

FIGURE 7.2 Tulga's task for deciding where to attend and act.

screen as blocks of differing heights and widths, then moved at constant horizontal velocity toward a vertical "deadline" at the right. Upon reaching the deadline, a block disappeared and there was no further opportunity to "work" it. Working any particular block meant holding a cursor on that block, which reduced the width of that block at a constant rate. There was but one cursor, while there were many tasks (blocks) on the screen, so the decision maker had to decide among them continually. Points were earned by completing a task block, that is, reducing its width to zero. The total reward for a task was indicated by the area of that block, the reward earned per unit time by the height of the block.

The human decision makers in this task did not have to allocate attention in the temporal order in which the task demands appeared or in the order in which their deadlines were anticipated. Instead, they could attend first to the task that had the highest payoff or took the least time, and/or could try to plan ahead a few moves so as to maximize points gained. The decision maker could quit working on a task and go back to it later, but there was a finite transition time required to move from one block to another. The paradigm was rich with experimental variables and dynamic decision hypotheses, and variants of it have been used by other experimenters. Tulga and Sheridan also derived an NP-complete dynamic programming algorithm to find optimal solutions for the multitask problem, based on scheduling theory (Pinedo, 2012).

The experimental results of Tulga and Sheridan suggest that subjects approached optimal behavior (as defined by the normative algorithm) when they had plenty of time to plan ahead. When there were more opportunities than the subject could possibly cope with, they tended to select the task with the highest payoff regardless of the time to deadline. This was evident because the optimization algorithm that best fit the subjects' data for this case discounted the future as the task inter-arrival rate increased. The subjects' corresponding subjective reports were that their sense of mental workload (Chapter 5, Section "Mental Workload") was greatest when by arduous planning they could barely keep up with all the tasks presented.

As the screen filled with more task demands and/or there was less time available before the deadline (as the "objective mental workload" grew), there came a point where the strategy switched from "plan ahead" to "put out fires and shed all other tasks." The latter meant doing whatever had to be done in a hurry while there was still time—and never mind that some very important long-term task could be worked on "ahead of time," because there might not be time to complete it. Interestingly, the investigators found that as "objective mental workload" continued to increase beyond this point where "plan ahead" had to be abandoned in favor of "do what has an immediate deadline," the subjective mental workload (as measured on a subjective rating scale) stopped increasing and actually decreased. A first-order quantitative hypothesis was that the mental workload is the disjunction (minimum) of utilized capacity (which kept rising up to saturation) and relative performance (which was nearly constant until capacity saturation, then diminished).

8

IMPLEMENTING AND EVALUATING THE ACTION

TIME TO MAKE A SELECTION

Hick (1952) proposed a model, based on Shannon's (1949) information measure H (see Chapter 5, Section Information Communication), for the time T it takes to choose which of several alternative movements, i, to make. The choice is based on the information in an immediately displayed signal calling for that move and the assumption that the move time itself is brief and constant,

$$H_{\text{choice}} = \Sigma_i p_i \log_2\left(\frac{1}{p_i}\right), \ T_{\text{choice}} = \alpha + \beta H_{\text{choice}},$$

where H_{choice} is information in bits, p_i is the probability of signal i, T_{choice} is the time required to choose, and α and β are scaling constants dependent on task conditions. α includes at minimum the base reaction time for making the slightest hand movement in response to a visual stimulus.

Modeling Human–System Interaction: Philosophical and Methodological Considerations, with Examples, First Edition. Thomas B. Sheridan.
© 2017 John Wiley & Sons, Inc. Published 2017 by John Wiley & Sons, Inc.

TIME TO MAKE AN ACCURATE MOVEMENT

Fitts (1954) also used the information measure for his model, in this case of the time required for making a discrete arm movement to a bounded location or target,

$$H_{move} = \log_2\left(\frac{2A}{B}\right), T_{move} = \alpha + \beta H_{move}$$

where (see Figure 8.1) H_{move} is information in bits (sometimes also called *index of difficulty*), A is the distance moved, B is the tolerance to within which the move must be made (for Fitts' experiment a tap between two lines), T is the task completion time, and α and β are again scaling constants, different for different conditions. Fitts probably did not realize what wide application the model would find. When applied to simple one-dimensional movements to a new position bounded within tolerances, the model has withstood the test of time and been robust over a wide range of A's, B's, and other task conditions such as bare-handed or movement via master–slave manipulator. It was successfully fitted to experimental data in a number of studies. However, like so many elegant models for human behavior, Fitts' model breaks down for more complex manipulations.

If a person must first make a choice and then move, a first-order model for time required will then be

$$T_{total} = T_{choice} + T_{move}.$$

Caution is suggested about expecting such simple assumptions of independence to be borne out by experimental data. Thompson (1977), in his

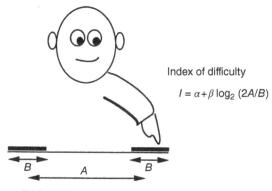

Index of difficulty

$I = \alpha + \beta \log_2 (2A/B)$

B A B

FIGURE 8.1 Fitts' index of difficulty test.

studies of manipulation, showed that the time required to mate one part to another was a function of the *degrees of constraint* (the number of positions and orientations that simultaneously have to correspond before the final mating could take place).

CONTINUOUS FEEDBACK CONTROL (SEE ALSO APPENDIX, SECTION "MATHEMATICS OF CONTINUOUS FEEDBACK CONTROL")

Continuous feedback control occurs where a sensed error in performing a task precipitates a corrective movement, much as the driver of a car continually senses error relative to the roadway course the driver wishes to follow and makes corrective movements on the steering wheel. Feedback control has a rich history. It has been apparent since antiquity that self-correction is inherent in nature. Especially since the Enlightenment, we have discovered self-correcting mechanisms in animals that focus the eyes, move the limbs, and so on.

The classical diagram for feedback control is shown in Figure 8.2, with the labels in this case relevant to a human operator as providing the control logic. The blocks represent input–output functional dynamic transformations such as differential equations (G_c for *controller*, G_p for controlled *process*) on the circulating signals, and the circles with the Greek sigma represent simple summation (subtraction when there is a minus sign on the input).

Think of the driver of a car. If there is any error e or discrepancy between a desired reference state r (the direction where one intends to go) and the visually observed actual direction y, the human operator adjusts the steering

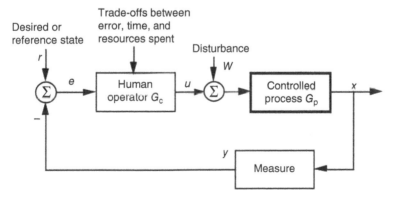

FIGURE 8.2 Classical feedback control system.

wheel to make a correction u. The car (in general a controlled process) responds accordingly, heading in direction x, which is measured visually, indicating a new sensed direction y. If there is any continuing error, the driver makes another correction. Because of twists and turns of the road (a continually changing r) as well as wind disturbance w, or imperfection in steering or vision, the driver must make further corrections, continuously trying to drive the error to zero. This is how any feedback control system works.

Self-correcting mechanisms are now common in everyday life. Thermostats regulate the temperature in our home heating systems. Float valves control the water level on our toilets after we flush. Automatic elevators take us to the desired floor. Autopilots steer our ships and aircraft. A well-developed engineering theory of automatic control implemented in computers enables great sophistication in today's automation. The mathematics in Appendix, Section "Mathematics of Continuous Feedback Control" summarizes the basic ideas of the feedback control model.

Feedback control models have been applied extensively to human–vehicle control. Of course, they have been applied to all sorts of other physical systems, and have also seen application to economics, marketing, management, and other fields.

In general, the controlled process is some complex dynamic electromechanical system best represented by a differential equation (essential because the variables continuously change with time). The general control problem is to make system output state x match desired state r despite disturbances w (wind buffeting, electrical or mechanical interference, etc., depending on context) and error or delay in measurement. The objective is to get a fast and smooth response to any changes in r and prevent instability. A system faces instability when the time lags that occur around the control loop add up to one half-cycle at any significant frequency (sine wave) component in the signals circulating around the loop, and the loop gain at that frequency exceeds one. When this occurs, what is meant to be negative feedback, that is intended to reduce error, becomes positive feedback, so that the error increases, potentially without bound. Thus the controller, be it human or electromechanical, must compensate to prevent instability.

Feedback control by human operators has an interesting history. During the 1950s, the U.S. government began a major effort to determine the equations characterizing pilots controlling aircraft. The reason was that in high-performance aircraft, the pilot's control characteristic must be made compatible with the aircraft dynamic equation, since it is the whole closed loop that determines control performance and stability. What became evident was that, within bounds, the human tends to compensate for the controlled process dynamics so as to make the entire loop behave properly. This is well specified in mathematics (McRuer and Jex, 1967; Sheridan and Ferrell, 1974), and was one of the first instances where engineers and psychologists came together to provide a sophisticated scientific solution to matching humans to machines.

The widely accepted model of human error tracking takes the form (in words)

System output $x = K$ (time integral of error e) (delayed by T),

where K is a gain coefficient and T is the human neuromuscular time delay of around 0.2 s.

Note that this defines the characteristic x/e for the *open loop* $G_c G_p$ and not that for the *closed loop* x/r. The important discovery, as noted before, was that the human controller G_c compensates for the controlled process G_p so that the combination stays stable and relatively constant.

This result is not so critical in aviation nowadays since, except for light private aircraft, autopilots have taken over and there is less need for human pilots to be direct-in-the-loop controllers.

The elementary mathematics of feedback control is presented in Appendix, Section "Mathematics of Continuous Feedback Control."

LOOKING AHEAD (PREVIEW CONTROL) (SEE ALSO APPENDIX, SECTION "MATHEMATICS OF PREVIEW CONTROL")

Unlike physical tracking systems that respond as best they can to continuous instantaneous errors between imposed reference inputs and system outputs, human controllers can usually look ahead to what the desired position or other reference input values will be before it is necessary to start responding. Automobile drivers can see "what's coming" on the road ahead; they do not base their control on seeing only the present lateral position or orientation of the car relative to the highway lane. Aircraft pilots can look ahead to the runway before landing or can fix their position on the map before increasing altitude to avoid a mountain. Power plant operators can look ahead to the procedural steps that must be taken up next.

In manned systems that do not provide preview, there is always a time lag in human response. Preview permits a response to be anticipated, prepared for, and initiated with no time lag and with some closer-to-optimal allocation of resources. Just as preview control is appropriate to direct manual control, it is also useful for the supervisor who must look ahead in vehicle control or other tasks and decide what goal to give to an automatic controller at each time step.

Given some assignment of the choices at each of a series of decision stages, and given the costs or rewards of taking any particular incremental decision at a particular stage, and assuming that costs or rewards are linearly independent, an optimal solution can be obtained either by dynamic programming (Sheridan, 1966) or by modern control theory (Tomizuka, 1975). Appendix, Section "Mathematics of Preview Control" illustrates an example of how dynamic programming might be applied in such a staged decision problem where the operator could preview and, hence, anticipate what obstacles or other costs lay in store at each future point in time. Human behavior can then be compared to this optimal behavior in experiments.

Reality dictates that the human operator cannot see infinitely far ahead, only a relatively short distance, and that the farther ahead one looks, the poorer the quality of that future information as relevant to decisions to be taken "now." The operator knows this and surely discounts information as a function of distance (time) ahead. Thus, preview models must include some preview-discounting function.

Seeing ahead and anticipating helps not only in lateral control of a vehicle but also in longitudinal control—speeding up on the straightaways and slowing down on the curves to stay behind lead vehicles. In multidiscrete task situations, this is equivalent to moving more quickly through the assigned tasks when they are easy and less quickly when they are hard. The term *self-pacing* may be applied in both continuous and discrete control situations.

DELAYED FEEDBACK

When there are significant telecommunication delays within the control loop, as in controlling a lunar vehicle from earth, continuous control produces instability, for reasons stated before. Ferrell (1965) simulated control of a lunar manipulator from earth (roughly 3 s round trip for a radio signal). He showed that humans naturally select a "move-and-wait-for-feedback" strategy that is quite predictable (Figure 8.3), being an approximate function of the delay plus a function of the Fitts' index of difficulty (Figure 8.1).

CONTROL BY CONTINUOUSLY UPDATING AN INTERNAL MODEL (SEE ALSO APPENDIX, SECTION "STEPPING THROUGH THE KALMAN FILTER SYSTEM")

Complex computer-controlled machines and systems all over the world are based on incorporating a model of the controlled process within the controller. In essence, it is a way to discover the essential features of a physical environment

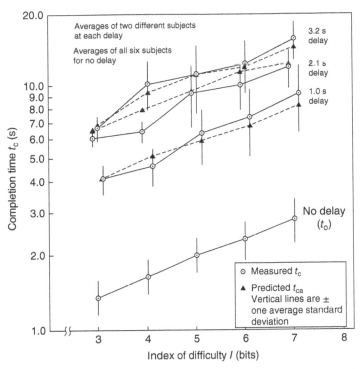

FIGURE 8.3 Ferrell (1965) results for time to make accurate positioning movements with delayed feedback.

one wishes to control, and to embody the properties of that environment (the external reality) in a computer-based "internal model." This approach has the advantage of allowing open loop ("dead reckoning") for short periods, while at the same time using feedback information as it is available. In many ways, it can be said to characterize how humans do subconscious tasks like walking, climbing stairs, and doing routine manual activities. Basing control only on feedback error has its limitations due to multiple demands on sensing, including delays and noise in measurement. Watching one's own feet can lead to stumbling and poor athletic performance. Kleinman et al. (1971) developed a widely cited "optimal control model" based on the internal model idea, which they applied to human control of aircraft, including the visual scanning and muscle neurodynamics. The internal model idea is described in detail in Appendix, Section "Stepping through the Kalman Filter System." In its engineering instantiation, the idea is attributed to Kalman (1960). Appendix, Section "Stepping through the Kalman Filter System" walks the reader through the way such a control system functions while sparing the reader the full mathematics.

EXPECTATION OF TEAM RESPONSE TIME

Experience has shown that human response times for a broad class of emergency responses (interventions) are distributed in log-normal fashion. Figure 8.4 shows an example for time to read and perform procedural steps in a large-break loss-of-cooling accident (LOCA) in a nuclear power plant (manned simulation exercise). The data are from Kozinski et al. (1982), where each data point is for a different team of reactor operators in a training exercise. Most of the responses occur within a very short time of the initiating event; a few require a long time. Clearly, the cumulative probability (that response will have been made before the time on the ordinate) will slope from the hypothetical origin at lower left (for quick responses) to 1.0 at upper right (for very long response times).

It may be difficult to understand Figure 8.5 because of how the axes are structured, but plotting the data this way results in a simple straight-line *log-normal* model of the data. The horizontal axis in this case is the cumulative fraction of test subject teams that responded within the time shown in the

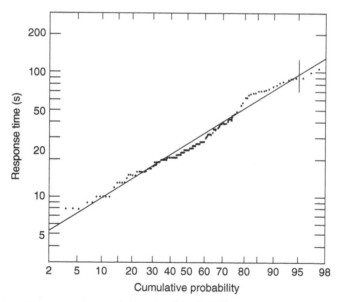

FIGURE 8.4 Response times of nuclear plant operator teams to properly respond to a major accident alarm. For the particular mathematical function used (log normal), using specialized graph paper (logarithm of response time on y-axis, Gaussian percentiles on x-axis) reduces that function to a straight line. The 95th percentile mark is seen to be roughly 100 s. Source: Kozinski et al. (1982). Reproduced with permission of U.S. Government Report.

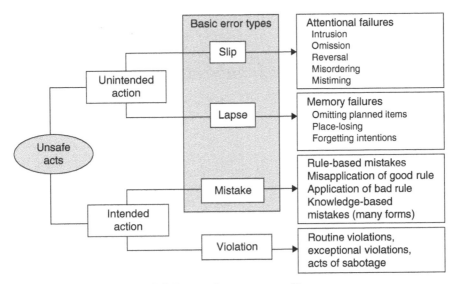

FIGURE 8.5 Reason's taxonomy of human error.

vertical axis (scaled logarithmically). This author, having been one of the experimenters, and having known that the particular distribution called log normal has been known to fit human response time data in simple experiments (Sheridan, 2013), tried that mathematical function and discovered a pretty good fit.

HUMAN ERROR

Cognitive engineering practitioners have long been interested in tallying and classifying human error. Errors can occur in sensing, memory, decision, or response. There can be errors of omission or errors of commission. Errors can be forced by external circumstances or just seem to occur randomly. The most popular classification is that by Reason (1991), shown in Figure 8.5, though there are others. Particularly important distinctions are between (1) *slips* (where a person intends the right action but because of inattention to the task did something unintended), which can fit into one of the categories to the right; (2) *lapses* of memory, resulting in one of the three error classes on the right; and (3) *mistakes* where one does what was intended, but it turns out to be wrong by general consensus. Intended violations are not normally called errors.

There are many potential causative factors. One is the lack of feedback in executing a task. Another is having an invalid mental model. Another is

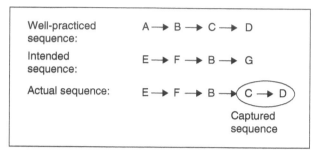

FIGURE 8.6 Capture error.

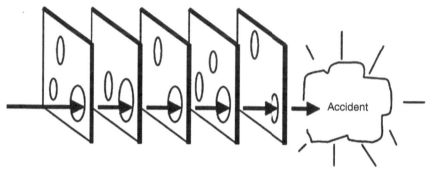

FIGURE 8.7 The Swiss Cheese model of accident occurrence as a result of penetrating multiple barriers. After Reason (1991).

mental or physical stress, and the resulting "narrowing" of situation awareness. An interesting class of error is known as the *capture error* (Figure 8.6). This occurs when a person deviates from a well-practiced sequence of actions (e.g., the standard drive from home to work), and on a rare occasion intends a different sequence that happens to include at least one of the steps in the well-practiced sequence (e.g., intended sequence in the diagram is a Saturday set of errands that includes *B*). Inadvertently, the Saturday driver ends up in the parking lot at work!

With respect to error prevention, there is general agreement that warnings are probably the least effective means of prevention, though lawyers love them for obvious reasons. Restricting exposure is effective if feasible. Usually, most effective is proper design of systems, based on thorough analysis of potential hazards as well as proper training.

A popular characterization of accidents, also attributed to Reason (1991), is the *Swiss Cheese model*, shown in Figure 8.7. The idea is that most accidents happen because multiple "barriers" are breached. These include the cognitive barriers such as proper training and intention and sufficient attention, as well as physical barriers built into the system (e.g., a working traffic light, working brakes, and a fastened seat belt). The Swiss Cheese model is similar to a much older metaphor called the *domino effect*.

However, there are those who claim that this model is deceptive, in that it implies that error and accident causality are one way. The point is that along the way (for any given slice of cheese), there are usually many opportunities for feedback and mitigation that preclude the situation getting closer to an accident of significant consequence. One popular model for this (Leveson, 2012) will be discussed in Chapter 11 in terms of its relevance for large-scale socio-technical systems.

9

HUMAN–AUTOMATION INTERACTION

HUMAN–AUTOMATION ALLOCATION

In 1951, Fitts published a list of "what men are better at" and what "machines are better at," sometimes known as the "Fitts MABA–MABA list" (Table 9.1). The list is obviously out of date because modern sensors and computers currently outpace people in several of the categories. There have been various international meetings on human–computer allocation, but consensus on a replacement for the Fitts list is elusive, perhaps because sensor and computer capabilities keep getting better.

To counter the then prevailing popular notion that systems were either automated or they were not, Sheridan and Verplank (1978) proposed what has come to be known as *levels of automation* (Table 9.2). It motivated a variety of reactions, including criticism that it did not provide guidance on its use in system design (which was never intended in the original paper). Indeed, a "levels" taxonomy can take many different forms and be multidimensional, and so on, and many versions have been proposed by different authors (e.g., Kaber and Endsley, 2003).

Parasuraman et al. (2000) proposed that the levels of automation can also be quite different for each of the four recognized stages of behavior for a human

Modeling Human–System Interaction: Philosophical and Methodological Considerations, with Examples,
First Edition. Thomas B. Sheridan.
© 2017 John Wiley & Sons, Inc. Published 2017 by John Wiley & Sons, Inc.

TABLE 9.1 Fitts' list

Men are better at
• Detecting small amounts of visual, auditory, or chemical energy
  ○ Perceiving patterns of light or sound
  ○ Improvising and using flexible procedures
  ○ Storing information for long periods of time and recalling appropriate parts
  ○ Reasoning inductively
  ○ Exercising judgment

Machines are better at
• Responding quickly to control signals
  ○ Applying great force smoothly and precisely
  ○ Storing information briefly, erasing it completely
  ○ Reasoning deductively

TABLE 9.2 The original levels of automation scale

1. The computer offers no assistance, human must do it all.
2. The computer offers a complete set of action alternatives, and
3. narrows the selection down to a few, or
4. suggests one, and
5. executes that suggestion if the human approves, or
6. allows a restricted time to veto before automatic execution, or
7. executes automatically, then necessarily informs the human, or
8. informs the human after execution only if he asks, or
9. informs the human after execution if it, the computer, decides to.
10. The computer decides and acts autonomously, ignoring the human.

Source: Sheridan and Verplank (1978).

performing a task, namely acquiring information, analyzing (and displaying) the information, deciding on action, and implementing (and evaluating) that action (Figure 9.1). These are the same stages that constitute Chapters 5–8.

Johnson et al. (2011) offer reasons why the original Sheridan and Verplank scale is insufficient for systems design purposes and does not account for interdependence factors.

SUPERVISORY CONTROL

Human supervisory control emerged as a trend in the 1960s (Ferrell and Sheridan, 1967; Sheridan and Johannsen, 1976; Sheridan, 1992a). Originally, it grew out of efforts to study whether and how human operators on earth

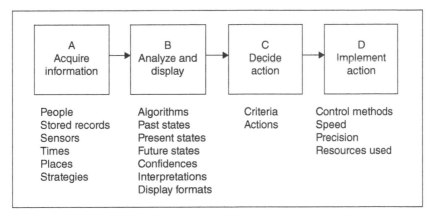

FIGURE 9.1 Four stages of human operator activity.

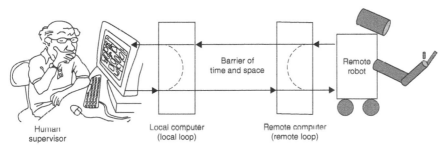

FIGURE 9.2 Supervisory control, as originally proposed for lunar rover operations (Ferrell and Sheridan, 1967).

could perform control tasks on the moon. It was already appreciated that control through a long-time delay was an invitation to instability, and Ferrell (1965), as described in Chapter 8, Section "Delayed Feedback," showed that human controllers could not magically overcome that physical constraint. Sensors, effectors, and computers were becoming inherent in lunar vehicles then being designed, such that the on-board computers could accomplish small segments of tasks automatically (without help from the ground). They did this by closing the control loop from sensors to actuators at the remote site. But the computers lacked the intelligence to comprehend the larger situation and know how to weave together the component task segments to accomplish the overall task and avoid catastrophe. Human supervisory control was an obvious solution.

Figure 9.2 illustrates the idea, where a computer local to the earth-based human would assist the human supervisor to display the latest (though delayed) information and to aid the human to program what to do next.

Packets of instructions and contingency information would be sent across the barrier of space (and time delay) to the lunar computer, which in turn would perform the elemental actions.

Human supervisory control of one or multiple remote computer-controlled operations is clearly analogous to that of a human supervisor in a factory or other industrial operation, where the subordinates can be other computers or can be other humans. For example in a nuclear power plant or large chemical processing plant, there may be hundreds or even thousands of control loops that may have to be supervised from a central control room and programmed/ monitored on very disparate time scales.

Accordingly, there are five supervisor tasks that can be identified: (1) planning the task(s) (and what is to be programmed); (2) teaching (programming the one or more remote computers); (3) initiating and monitoring automatic control for failure or abnormality (different at each site); (4) intervening and reprogramming in the event of failures, or termination of programmed tasks; and (5) recording results and learning from experience. Figure 9.3 diagrams these functions.

In consideration of the long-term trends in supervisory control with respect to various types of tasks, Figure 9.4 suggests a two-dimensional relationship, with degree of automation on the horizontal axis and task entropy on the vertical. "Task entropy" is a way of characterizing *disorder to be overcome*, unpredictability, or similar challenges posed by the given task. Composing symphonies, managing large institutions, or raising families might be said to be dignified human work, opposite on the task entropy scale to a human performing completely predictable and routine menial labor. Automation performing fully predictable tasks can be called "clockwork." Matching human capability by machine in sophisticated tasks might be called the "ultimate robot." The slant line suggests a "line of progress," which raises ethical and moral questions of how far to go in automation, which has both advantages and serious implications for jobs and other aspects of being human.

TRADING AND SHARING

The roles of the computer in supervisory control can be classified according to how much task-load is carried compared to what the human operator alone can carry. As suggested in Figure 9.5, the computer can *extend* the human's capabilities beyond what the human can achieve alone; it can partially *relieve* the human, making the human job easier; it can *back up* the operator in case the human meets capability limits; or it can *replace* the human completely. Extending and relieving are examples of *sharing* control. More specifically,

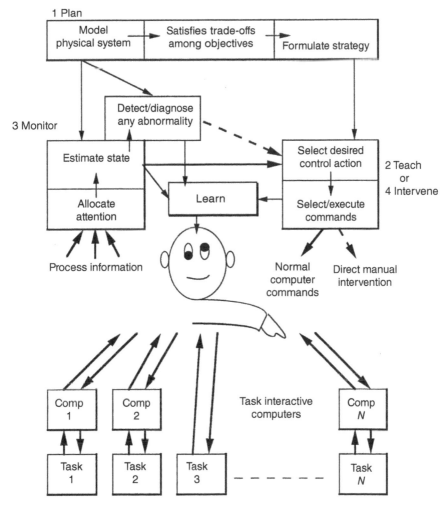

FIGURE 9.3 Functions of the supervisor in relation to elements of the local human-interactive computer (Figure 9.2) and multiple remote task-interactive computers.

sharing control means that the human and the computer control different aspects of the system at the same time. Backing-up or replacing can also be called *trading* control. Trading control means that either the human or the computer turns over control to the other (or the latter seizes it). In sharing control, the main issue is: *which tasks* should be assigned to the human and which to the computer? In trading control, the main issue is: *when* (e.g., conditions with respect to time, completion of assigned task, error or deviation from nominal plan, expectations of the other's readiness) should control be handed over willfully and when should it be seized back? There are also

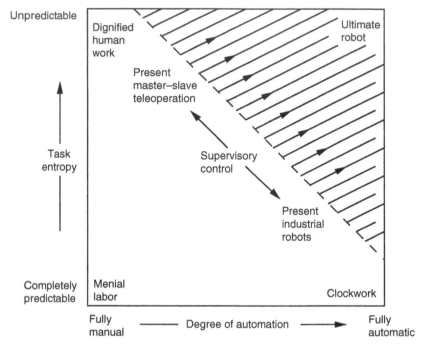

FIGURE 9.4 Supervisory control in relation to degree of automation and task entropy.

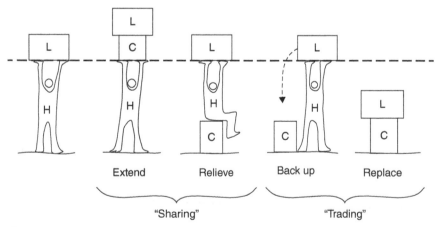

FIGURE 9.5 Distinctions with and between trading and sharing control. Source: Sheridan and Verplank (1978).

forms of cooperative control where control is initiated by one agent (human or computer) and the other trims or refines it. None of these modes is purely hierarchical, with one agent always superior and the other always subservient (i.e., supervisory control as defined earlier).

Rouse (1977) utilized a queuing theory approach to model whether from moment to moment a task should be assigned to a computer or the operator should do it independently. The allocation criterion was to minimize service time under cost constraints. Results suggested that human–computer "misunderstandings" of one another degraded efficiency more than limited computer speed. In a related flight simulation study, Chu and Rouse (1979) had a computer perform those tasks that had waited in the queue beyond a certain time. Chu et al. (1980) extended this idea to have the computer learn the pilot's priorities and later make suggestions when the pilot is under stress.

Moray et al. (1982) experimented with an arrangement wherein the operator, faced with four independent control loops, could either monitor the systems and control them manually or leave them to automatic control. The aim was to keep all systems within fixed limits. They found that the initial attention of relatively unpracticed operators was almost equally divided between systems, despite markedly different time constants to reach critical limits.

ADAPTIVE/ADAPTABLE CONTROL

The term *adaptive control* refers to a computer automatically adjusting controller parameters as a function of the measurement. For example, an airplane automatically makes compensating gain adjustments with measured altitude as the air gets thinner and the movement of the flight control surfaces has less effect. *Adaptable control*, in contrast, refers to what a human supervisor does in adjusting system parameters, for whatever reason.

Figure 9.6 details several different modes of adaptable control, various things the human supervisor can do to adjust parameters of a control system. The variables r, u, x, and y in the diagram correspond to reference input, controller output, controlled process output (system state), and measurement of state, the same as in the standard feedback control diagram mentioned earlier. u^* and x^* are, respectively, the control variable and process output of a simulation that the supervisor might use offline to help in planning what actions to take with the online system. For example, in air traffic control, the controller has simulation capability to test whether extrapolation of current aircraft trajectories will conflict, or whether an aircraft can safely navigate through a given weather pattern.

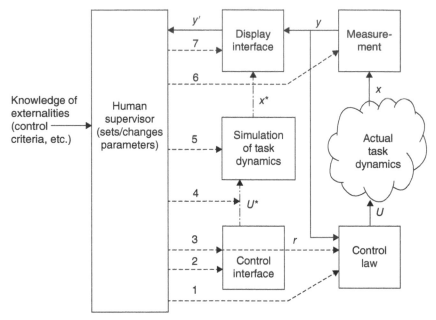

FIGURE 9.6 Adaptable control (from Sheridan, 2011).

The numbers in the diagram represent (1) control law adjustment; (2) adjustment of the control interface, such as the level of force feedback in a hand controller; (3) modification of the input r which the automation is tracking; (4) resetting the input u^* to an offline simulation, such as "what would happen if u^* was the input"; (5) modifying the simulation dynamics for the given input; (6) changing the type of measurement, for example, making it more or less sensitive, or measuring a different variable; and (7) changing the display interface, for example, from analog to a digital display, or to one having a different scale. Clearly, an operating control system has many properties that can potentially be altered by a human supervisor.

MODEL-BASED FAILURE DETECTION

The idea here is to use a reference model that represents the way the *subsystems* of the operating system (A, B, and C in Figure 9.7) *should* behave (based on prior verification tests), and then to make comparisons to what is actually happening. The states of variables at various points within the actual operating system (as well as at the final output) are fed to *subsystem models* (shaded blocks) and the outputs of those subsystem models are compared with corresponding states of the actual operating system continuously in real time.

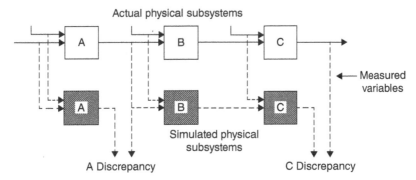

FIGURE 9.7 Model-based failure detection.

If something fails (or significantly deviates from normality) in the actual system, there will be deviations from the models that become evident in the corresponding comparisons.

Sheridan (1981) (see also Tsach et al., 1982) experimented with this technique in conjunction with an actual operating fossil power plant. Effort and flow covariables (e.g., fluid pressure and flow, shaft torque and speed, and electrical voltage and current, all of which pairs are called "power bonds") were measured at various key points in the plant, at least one such pair for every state variable. The segmentation for component models occurred at these points, with the effort forcing the submodel on one side and the flow forcing the submodel on the other side (as determined by causality in the real system). The model covariables were then compared to the actual measured covariables, and any significant discrepancy indicated a failure somewhere, as located by the submodel from which the discrepant covariable was an output. By cuts and comparisons at different places in the system, a failure could be located to a reasonable degree of precision. Though the energy covariables afford one basis for model disaggregation, causality is the only necessary condition. In one instance, the researchers notified the actual plant operators of an incipient failure.

10

MENTAL MODELS

WHAT IS A MENTAL MODEL?

At this point, it is important to consider what have been called *mental models*. These are not what we have been calling scientific models (i.e., derived from openly available and measurable evidence, and rendered in words, graphics, or mathematics so that anyone can make the same observation to try to get the same result). In contrast, mental models are hypothetical renderings of objects or events in the brain of a person. The events from which they are inferred are private to the subject, but an experimenter can repeat an experiment to see if the subject says or does actions that lead the experimenter to the same inference. Depending on context, a mental model can incorporate notions of sequence or movement in time, probability, value, contingency (what if), explanation, and are often visual. They need not be limited to empirical facts, but can be imaginings.

There is no way to know with certainty what anyone's mental model actually is. One can only make inferences from what the subjects say, how they act, what choices they make in a contrived experiment, or what they put on paper, and trusting that they are telling the truth about their thoughts. Clever experiments comparing skilled task performance with the subject's

Modeling Human–System Interaction: Philosophical and Methodological Considerations, with Examples, First Edition. Thomas B. Sheridan.
© 2017 John Wiley & Sons, Inc. Published 2017 by John Wiley & Sons, Inc.

verbalization of what they were thinking (Sanderson, 1989) have allowed inferences of mental models. Presumably, scientific models have evolved from such studies of the mental modeling process, refined over time until the investigator is comfortable communicating the findings openly in print or to other people. In any case, as previously noted, mental events *per se* do not meet the observability criterion discussed in Chapter 2.

Mental models have practical uses, insofar as they can be trusted. They reveal preferences and misunderstandings in using equipment or interacting with other people. Teachers and trainers want the learner/trainee to develop the correct mental model, and the mental model can be said to reveal a person's level of understanding or competence.

Does the notion of mental model mean something other than knowledge in general? The term is almost always used to mean something more specific: an inferred representation of the structure, function, and causality properties of a physical subsystem or task, evolved over time, and stored in long-term memory. That representation can be qualitative or quantitative. A utility function derived by the Von Neumann process (Appendix, Section "Mathematics of Human Judgment of Utility") is an elegant quantitative mental model, elicited by a very simple iterative decision process.

Given any scientific model (i.e., one based on observable events), a human user must have a mental model of that scientific model. The user of the particular scientific model must understand it and commit its essentials to memory so as to integrate it into the user's own cognitive process and potential behavior. So, questions of language familiarity, clarity, adherence to accepted norms of expression, and understanding arise. In other words, the model user must develop an own mental model corresponding to the scientific model.

BACKGROUND OF RESEARCH ON MENTAL MODELS

The notion of a mental model has existed in the literature of psychology for at least 65 years. Craik (1947) used the term to mean a "model of the world in the human nervous system." He suggested in that the mind constructs "small-scale models" of reality that it uses to anticipate events.

Other definitions (from Wikipedia) are as follows: "explanation of someone's thought process about how something works in the real world"; "a representation of the surrounding world, the relationships between its various parts, and a person's intuitive perception about his or her own acts and their consequences"; and "an internal symbol or representation of external reality, hypothesized to play a major role in cognition, reasoning, and decision-making."

The image of the world around us, which we carry in our head, is just a model. Nobody can imagine all of what goes on in his surround. He has only selected concepts, and relationships between them, and uses those to represent some real but bounded system of concern. Mental models can help shape behavior and set an approach to solving problems (akin to a personal algorithm) and doing tasks.

The fashion in psychology encouraged by Skinner (1938) in his *Behavior of Organisms* was that there is nothing directly observable about peoples' thoughts, so the only admissible scientific evidence about people is their overt behavior. But all that changed as the computer provided a new metaphor, and now what we call cognitive science came into fashion. The computer metaphor does not demand that the structure of the cognitive model be physically like brain physiology, only that it functions as though it could be simulated by a computer.

Many researchers have picked up on the mental model term, but whether they mean exactly the same thing is open to question, since the context of their use of the term varies widely. Johnson-Laird, for example, infers subjects' mental models in doing logical reasoning in solving simple puzzles. Gentner and Stevens (1983) use the term in their studies of how people infer causality. Schiele and Green (1990) talk about mental models of how people interact with computers. Jagacinski and Miller (1978) refer to mental models of human operators of control systems. Endsley (1995a,b) writes about mental models as related to a person's "situation awareness," defining that term to be "the perception of environmental elements with respect to time or space, the comprehension of their meaning, and the projection of their status after some variable has changed, such as time, or some other variable, such as a predetermined event."

Norman (1983) has distinguished three different uses of the term, not with respect to the physical task contexts such as those just mentioned, but rather with respect to different models in the heads of the designer (of the thing or task), the operator or user (who interacts with the thing or performs the task), or the researcher (who studies the people either designing or performing the task). Indeed, there is a sequence of mental models that one might consider that are potentially different. Adding to the these three are the models in the heads of authors (like me) who review research, and the models in the heads of the readers (like you) who read what those authors write about the researchers who study the performers of the things and tasks the designers designed. Fleas have little fleas—almost *ad infinitum*.

Rouse and Morris (1986) review how different domains define and use the term "mental model" and various issues that arise, including accessibility, forms of content representation, cue utilization, and instructional issues.

What follows are brief expositions of four very different kinds of mental models: (1) Anderson's Adaptive Character of Thought (ACT-R), (2) Moray's characterization of a mental model of structure and causation in terms of graph theory, (3) Yufik's model of how the mental model forms in the brain, and (4) Pinet's discussion of the aircraft pilot's mental model for decision-making under extreme time stress.

ACT-R

ACT-R is a widely used theory and cognitive modeling tool originally developed by Anderson (1990) at Carnegie-Mellon University. ACT-R has been used to create models in domains such as learning and memory, problem solving and decision-making, language and communication, perception and attention, cognitive development, and individual differences. Beside its applications in cognitive psychology, ACT-R has also been used in human–computer interaction to produce user models that can assess different computer interfaces, education (cognitive tutoring systems) to "guess" the difficulties that students may have and provide focused help, computer-generated forces to provide cognitive agents that inhabit training environments, and neuropsychology (to interpret fMRI data). Closer to cognitive engineering, it is being applied to modeling aircraft piloting and highway vehicle control. There is open-source ACT-R software available (http://act.psy.cmu) that runs on any platform with a compatible LISP compiler.

The different modules of ACT-R are portrayed in Figure 10.1. A declarative memory module stores facts about the world the authors call *chunks*. The declarative memory module also includes facts about the current task goal. A visual module (and/or modules for other human senses such as hearing and touch) identifies objects in the visual field or environment. A motor module specifies how the human limbs are sequenced and timed to produce necessary movements such as key presses or grasping of objects. The "world" with which the vision and motor modules interact is necessarily simulated. All three of the aforementioned modules have buffers associated with them. The goal buffer keeps track of which step is current in progress toward the goal.

The procedural memory module stores *if-then* production rules (*productions*). The pattern matching module matches the appropriate rule to apply to the next step based on the *chunk* information in the declarative memory, vision (sensing), and motor modules. Only one step can be executed at a time. If multiple productions match the joint contents of the buffers, there are

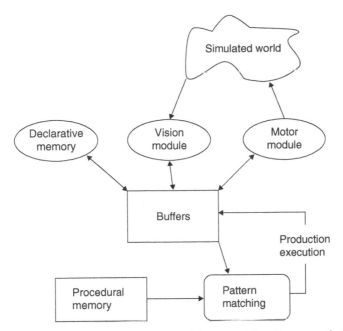

FIGURE 10.1 The ACT-R cognitive architecture (after Byrne et al., 2008).

conflict resolution rules that determine which production is allowed to fire. Firing (execution of an *if-then* production) occurs about every 50 ms, changing the contents of at least one buffer. Each chunk has associated with it some statistical information about frequency, recency of access, and relationship to the current context that can either be input by the user or learned through ACT-R operation.

ACT-R does not contain any advanced machine vision or machine learning in an artificial intelligence (AI) sense that would allow it to recognize objects in the environment. It needs explicit descriptions of such objects to be programmed in. This is why ACT-R is a model that generalizes in the same way that any computer program generalizes. It requires the user to input many free parameters that have to be explicitly tailored to fit any empirical data set, whereas most of the other models described in this book, though much simpler and narrower in their application, require fewer parameters to achieve a fit.

ACT-R has motivated other authors to devise similar computer-based (non-analytic) models. NASA sponsored a formal model comparison project to ascertain which and how well different such models could characterize pilots performing aircraft taxi operations. They included IMPRINT (Archer and Adkins, 1999) that combined high-level network modeling with the cognitive features of ACT-R; D-OMAR (Deutsch, 1998), an event-based simulator designed to facilitate multitasking behavior; (Wickens, 2002) "salience-effort-expectancy-value" (SEEV) framework described in Chapter 5; Endsley's (1995a,b) situation awareness model, also described in Chapter 5; and an evolving NASA model called "Air MIDAS" that started as a cognitive workload model but later took on perception, motor, and domain knowledge features similar to ACT-R. This joint evaluation effort constitutes a book by Foyle and Hooey (2008).

LATTICE CHARACTERIZATION OF A MENTAL MODEL

Moray (1997) proposes the use of lattice theory (Birkhoff, 1948) to represent mental models. He asserts that this particular way of mapping properties of some target objects or events efficiently captures the most important features of mental models.

A lattice (Figure 10.2 shows two examples) is a partially ordered set of elements (which become nodes on the lattice graph). Any two nodes are ordered by a diadic relation "≤" or "≥" (corresponding to "lower than" or

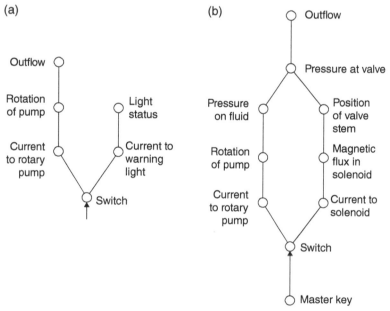

FIGURE 10.2 An example of Moray's 1990 lattice model of the operation of a pump: (a) causality relations and (b) purpose relations.

"higher than" in the lattice), which can be assigned any kind of meaning, such as "causation." Figure 10.2a might mean that the switch causes the flow of current to both the pump and the solenoid, and these in turn have other causal effects upward in the lattice. Nodes on the left branch are not causally related to those on the right branch until the cooccurrence of both pressure on fluid and position of valve stem cause pressure at the valve.

The diadic relations in Figure 10.2b can also be considered to mean "purpose"; that is, the purpose of the switch is to turn on current to the two nodes above it. However, purpose can also be thought of as levels of aggregation and abstraction. Thus, in Figure 10.2b, the switch is considered part of each of two pump systems, one of which is part of a normal pump system and one of which are part of an emergency pump system, both pump systems being part of the cooling system, the highest level of abstraction represented here.

Moray suggests that there are four criteria by which human operators map elements of the actual world into the same element or different elements at the same level of the abstract representation (formal lattice): (1) *physical resemblance* between elements, (2) *correlation* of the *behaviors* of elements, (3) *causal relation* between elements, and (4) *common purpose* among elements. The mental lattice is likely to be different for different criteria which the operator deems appropriate, even though the physical elements to be cognitively mapped may be the same. In discussing the meaning of such cognitive mappings, Moray suggests that the lattice may support more than static facts; it may embody rules for action.

Moray also makes the point that which of the (infinite set of) properties of the target things or events included in the mental model depends on the person's intentions in using the model. He claims that these uses form a hierarchy, from (1) goals, means, and ends (at a highest level); (2) general functions; (3) specific functions; to (4) physical form (at the lowest level in the hierarchy). This accords with the Rasmussen hierarchy of abstraction (1983) discussed in Chapter 3.

The lattice or diadic graph has been discussed here in relation to mental modeling. It has also been applied for various other purposes. For example, graph models can be used by groups of people to analyze their goals, by organizations to analyze their structures, and by manufacturers to analyze the time precedence of their manufacturing processes. Given the diadic relation between every possible pair of elements in a system (the *reachability matrix*, for example, the relative level of goals, the relative power in an organization, and the relative time precedence of two manufacturing operations), one can determine the relation of adjacent elements on the graph (the *adjacency matrix*), and thus the whole graph can be drawn.

NEURONAL PACKET NETWORK AS A MODEL OF UNDERSTANDING

Yufik (2013) has developed a model of neuronal clusters or "packets" that are associated with sensing and comprehending the world (understanding).

The idea is that cognitive tasks mobilize neurons to form packets that are associated due to cofiring (cooccurrence of features of the stimulus), resulting first in links (like A–B and C–D in the left-hand box of Figure 10.3), and eventually many interneuronal links (center). When the strength of links between certain neurons (internal weights W^{int}) exceeds those connecting to other neurons (external weights W^{ext}), packets form (right-hand figure). The latter is called a virtual associative network (VAN).

Changing stimuli and consequent firing patterns destabilize packets, causing expansion, contraction, merger, or dissolution. Packet stability is determined by the ratio σ of summary strength of internal links pulling neurons together to the summary strength of links that pull them apart (and may eventually dissolve them).

$$\sigma = \frac{\sum_{i \in R_K} W_i^{int}}{\sum_{j \in R_K} W_j^{ext}} \rightarrow max.$$

Behavior, in Yufik's model, is represented by activation of the packets that change their sensitivity to stimulus features but do not cause the packets to dissolve (thus preserving their individual identity). Each neuron can be said to be represented by a vector in the (very large) space of stimulus features, and a packet characterized by a vector sum of the component neuronal vectors.

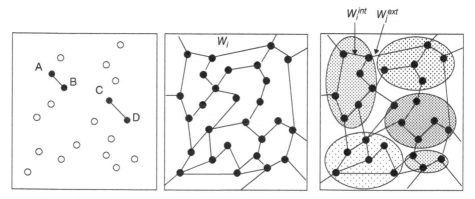

FIGURE 10.3 Formation of neuronal packets in Yufik's model of understanding.

Activation is equivalent to changes in the packet firing pattern, and thus changes in the packet vector (which means rotation in feature space). Yufik's claim is that mental modeling the process of "understanding" boils down to coordinated rotation of packet vectors. Consciousness, he asserts, is the experience of this process.

MODELING OF AIRCRAFT PILOT DECISION-MAKING UNDER TIME STRESS

Jean Pinet (2016), a prominent test pilot and pilot training official with a Ph.D. in psychology, developed an elaborate mental model using many of the elements of the *if-then* rule-based modeling popularized by Johnson-Laird (1983). Pinet's model relates to the airline pilot's cognitive activity in coping with high stress situations that often confront pilots with only seconds to spare.

One must respect airline pilots for the amazing skill they have acquired. We all know how to drive a car and operate the steering wheel and pedals to maneuver a car at speed and sometimes within a few feet of other cars that are stationary or moving in the opposite direction. Some of us know how to sail boats using a tiller and a few lines to tack upwind and come about with fickle wind gusts. But the airline pilot! To have acquired a mental model of how to operate multiple controls, some embodying a degree of automation, to regulate thrust, speed, attitude, altitude, and heading—all interacting in complex non-linear ways in three dimensions of space and time as governed by the physical constraints of aerodynamics and control—what is going on in the pilot's head?

Pinet develops a mental model of how pilots think under extreme time stress, down to a few seconds in some cases, and he applies that model to a variety of flight cases. The elements of the model include a number of notions from cognitive science: (1) an operating system, (2) agents (that perform specific *if-then-else* functions), (3) a differentiation between long- and short-term memory, and (4) a mechanism by which agents are retrieved from long-term memory to perform certain cognitive functions to interact with short-term memory. His descriptions of how the model works in each case are inter-woven with the explanations of the known physics of aircraft aerodynamics and control that the pilot must surely have coded into his memory. The model is couched in terms of verbal descriptions within scenarios. The scenarios are categories of situations a pilot encounters where abnormal events can arise (takeoff roll, climb to altitude, encounters with extreme turbulence, trajectory conflicts with other aircraft, descent and timing relative to other aircraft in the pattern, landing and go-around). For each scenario, there are descriptions of what the pilot sees, draws on memory to assess, decides and acts, referring in each case to the panoply of hypothesized cognitive functions and agents. So the set of *if-then-else* rules becomes very large.

I must admire this effort as an in-depth look at what aircraft piloting is all about. As a very inexperienced private pilot myself, I am very aware of the complexities and difficulties of mentally integrating all of what Pinet discusses. I would regard Pinet's model and his description of its use as a valuable tool for the nonpilot to get an idea of what piloting entails, and for the pilot to calibrate himself. Pinet's model is a valiant effort to frame a mental model for a very sophisticated type of task.

With regard to the latter, however, one might question whether such complex mental model as Pinet proposes is acceptable to the scientist who demands observation and model validation. This is the cloud that still hangs over mental modeling in general.

MUTUAL COMPATIBILITY OF MENTAL, DISPLAY, CONTROL, AND COMPUTER MODELS

Cognitive engineers have long recognized the need for a human operator to have an accurate mental model of the task being performed. They also subscribe to the importance of display-control compatibility, having what is seen on a display and what actions need to be taken at the control interface being compatible in geometric and logical form. As computer models increasingly play a role inside the system to allow for offline simulation (as was discussed in Chapter 9, Section Adaptive/Adaptable Control) or online control (as with the Kalman filter), it is critical that all four types of model be mutually compatible, as well as correspond to the real world, as suggested in Figure 10.4.

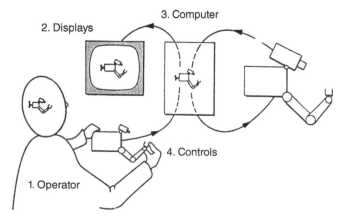

FIGURE 10.4 Multiple model representations in teleoperation. Source: Sheridan and Verplank (1978).

11

CAN COGNITIVE ENGINEERING MODELING CONTRIBUTE TO MODELING LARGE-SCALE SOCIO-TECHNICAL SYSTEMS?

BASIC QUESTIONS

In this chapter, we confront some larger questions concerning modeling in cognitive engineering or human–system interaction, and its relation to the "bigger picture":

1. *Scaling up:* Can our *micro-models* (of one or at most a few people interacting with their environments) contribute to modeling at a macro-level of large-scale societal issues? Can our micro-models "scale up" or provide insights to needed macro-models? Can some macro-models be constituted as the aggregate of micro-models, over time and across a variety of human–system endeavors, much as thermodynamics or other fields of physical interaction allows for such aggregation?

2. *Which models?* Which of our micro-models are best suited for such tasks? Can the cognitive engineering models simply be retooled to be fitted with different empirical data such as are now used in economics, politics, population, or weather studies? Will only those of our micro-models that are quantitative be useful, or will some of

Modeling Human–System Interaction: Philosophical and Methodological Considerations, with Examples, First Edition. Thomas B. Sheridan.
© 2017 John Wiley & Sons, Inc. Published 2017 by John Wiley & Sons, Inc.

our qualitative models also have potential for dealing with societal issues? Which societal issues? What are the criteria of usefulness or of optimization?

3. *Broadening human participation and interaction with the models beyond the researchers and policy-makers:* If human–system modeling, either quantitative or qualitative, can be retooled, is this something that should remain as a topic limited to graduate education and professional practice? Or would it make sense for it to be introduced as part of physical science, psychology, or other social studies at secondary education levels to motivate students to see social and physical phenomena in broader perspective of cause and effect? Will it accommodate easily to growing online computerized education (massive, open, online computer systems or MOOCs)? Can computer graphics and virtual reality (VR) help convey the implications of "what would happen IF" models? To what extent and in what way can the lay public get involved in a meaningful way? Can "crowd sourcing" through social networks provide needed data and subjective preferences as inputs to such models in real time? Surely there will be many human factors considerations.

WHAT LARGE-SCALE SOCIAL SYSTEMS ARE WE TALKING ABOUT?

Unavoidably, there are more questions than answers relative to the aforementioned propositions. Some examples of social systems that are calling for models are as follows:

- Health care
- Climate change and its long-term threats
- Air and ground vehicle transportation
- Automation, job displacement, and work satisfaction
- Inequity between the rich and poor
- Privacy–security trade-offs
- Population growth and supply of food, water, and other necessities of life
- Governance

Problems in these areas are highly interdependent, but we will try to comment on them separately.

Health care

The cognitive engineering community has been working at the micro-level in some of these areas for years. For example, in health care, nursing station alarms, surgical amphitheater layout, design of infusion pumps, nurse/physician data entry, and many aspects of patient safety have been actively investigated by cognitive engineering researchers. But cognitive engineers seem not to be much involved in the greater challenges of making health care cheaper and more efficient, offloading traditional doctor tasks to nurses or lower paid specialists, use of computer simulation in training medical personnel, and understanding the role of affective caring versus purely functional task performance.

There are major issues of how medical specialists allocate their time among patients. For example, see Sheridan (1970b) for a discussion of the problem of allocating "personal presence," and the degree to which telemedicine can improve health care in the trade-off with having to be "hands on." Can cognitive engineering modelers contribute to improving health care availability and efficiency in these broader contexts?

Climate change and its long-term threats

How we use energy is both a personal and a community decision. Energy conservation is an obvious need, but somehow we take energy demand as a given. We could save much energy by heating and cooling our bodies rather than our buildings, and by interacting via telecommunication and telerobotics rather than by travel. Why are we not moving in that direction? Understanding how individuals decide and the trade-offs with money, time, and convenience seem to be an obvious issue for cognitive engineers to research and model, yet I am not aware of much effort by them. Energy demands will probably militate in the direction of smaller homes in urban areas, better mass-produced and more available housing for the poor, especially in less developed countries, and rethinking how we live. Cognitive engineering has not much contributed to architecture and urban planning, but could and should.

Air and ground vehicle transportation

Everyone knows that urban traffic jams and waiting at airports are increasingly frustrating. Further, transportation is among the largest consumer of energy. Advances in telerobotics and the "internet of things" permit us to exercise control of systems from anywhere using cell phones or other body-worn devices. Advances in telecommunications permit us to attend meetings with other people anywhere. Combine the latter with

VR headsets and we can position our own bodies as avatars in those meetings for improved social interaction. So, telework and telesocial interactions become easier. However, such technology inhibits shaking hands or other personal interactions. These brave-new-world technologies would certainly save energy, but do they compromise what it means to be human? What are the trade-offs? Cognitive engineering models and research would seem to be called for.

Automation, job displacement, and work satisfaction

The standard assertion about robotics and automation used to be that it not only made production more efficient, and was indispensible for certain kinds of production (e.g., chip manufacture), but it also ultimately resulted in more employment. Workers could be replaced in low-level manual jobs but would move to higher-level jobs where they become supervisors of automation (as discussed in Chapter 9). Maybe some blue collar workers could not or would not take on white collar work. As computers and artificial intelligence (AI) assume more intermediate level desk jobs, there are further worries. How essential is "work" and what kinds of work are essential to living satisfying lives? Psychiatrist Erich Fromm popularized the importance of what he called "escape from freedom" and the need for a "productive orientation" (Fromm, 1941). Will serving each other as human beings largely replace attending to machines and the physical environment? It is a simple fact that cognitive engineering as we know it today has contributed relatively little to the rapidly growing field of human–robot interaction (Sheridan, 2016). Models in cognitive engineering often assume that "system performance" is the ultimate criterion to be maximized, but do we reach a point where human well-being trumps "system performance" (and what are the subtle differences)?

Inequity between the rich and the poor

This a shameful situation we all recognize. Considering only the United States (which we have some political control over), in a capitalistic society there are difficult questions regarding how to rectify the difference. Taxes? Education? As suggested in Chapter 5, Section "Logarithmic-Like Psychophysical Scales" and Chapter 6, Section "Valuation/Utility," both basic senses and utility judgments of people scale logarithmically relative to the physical variables. Insofar as valuation of dollar assets and happiness works that way, it is obvious that if everyone had the same dollar assets, the net utility (happiness?) would increase dramatically. What are the barriers (laws, cultural constraints, etc.) working against such an idyllic outcome? I do believe cognitive engineering modelers can play an important role in clarifying the implications of this fact (in addition to economists, sociologists, political scientists, etc.)

Privacy–security trade-offs

In recent years, the threats of terrorism, "cyber hacking," and so on, have added to our national insecurity beyond the traditional reasons for having a standing military apparatus. Our nation now has multiple security agencies, some military, and some civilian. Large amounts of taxpayer dollars are spent on information security of both organizations and individual persons. Nominally, in order for security agencies to do their jobs, they need to access telephone and Internet transmissions of all kinds. Yet, this very access is a threat to our privacy, a fundamental constitutional right. That there are trade-offs is clear, at least to policy-makers, but maybe not so much to the general public, who demand both security and privacy. Where to draw in line in terms of policy is not evident, nor is it clear how best to help the public understand these trade-offs. It seems that cognitive engineering modeling and research could play a very constructive role here.

Population growth and supply of food, water, and other essentials

On a worldwide basis, and over the long term, population growth is clearly one of the greatest threats to the continuation of our species. In the developed nations, population growth is tapering off, and China has now reversed its severe one-child policy (since there will be relatively too few young and working-class people as the general population ages). However, some nations in Africa and Asia that are least developed have the highest (and unsustainable) growth in population. Combine with these facts that progress in health is allowing people everywhere to live longer on average, creating a much larger fraction of elderly than in the past. This seems like progress, but is it? Is longevity a threat? Add to this the facts of increasing shortages of potable water in arid regions exacerbated by climate change, and we have major interactions that we hardly understand. There have been world models that have made dire predictions (e.g., the Club of Rome Limits to Growth studies) that have proved incorrect, so far. Nevertheless, the modeling challenge in this broad category of issues persists. Climate change, technology, economics, education, family planning, and many other variables interact. As with the other issues discussed before, there are many trade-offs. I do believe there is a challenge to cognitive engineering modelers to engage these problems, especially in terms of decision-making practices of individual citizens.

Governance

The existence of the Internet, social media, and the widespread availability of cell phones, laptops, TV, and other electronic devices has huge implications for governance. A century ago, the machinery of government interaction with the people was constrained to the print medium and time constants of days or weeks for communication from leaders down to the governed. Today, such communication is virtually instantaneous, and there is little or no limit on how much information can be distributed in either direction. It has been both a boon to democracy and a curse, in the sense that demagogues and disinformation can thrive. In a highly connected world, trade seems like a generally good idea, but it leaves out a certain fraction of the labor force in rich countries when labor is cheaper in other countries. To what degree should governmental policy decisions be made by experts, and to what extent made by the broader population of those affected? Currently, the United States has multiple layers of government (village, county, state, national) with much redundancy (and confusion) with respect to regulation, security, education, and administrative functions. With new technology, could governance be made simpler and more efficient?

There are other large-scale societal systems that could be added to the list. They all involve people interacting with technology. The question of how human factors/cognitive engineering can help cope is certainly not novel, and others have dealt in their own ways with that question, for example, see Nickerson (1992).

WHAT MODELS?

In considering which models of all those reviewed in the previous chapters are most appropriate to the gargantuan task suggested here, I can only offer some general thoughts.

Assessing every model relative to every societal problem is far too difficult.

Context dependence of models: Some models are more context dependent; some are less so. The models of feedback control and information communication were originally borrowed from engineering as noted before, and applied to human factors problems. Theories of games and decision-making are borrowed from economics. They are very general and context independent. Certain other models are devised for particular human work situations or technologies, such as piloting aircraft, or VR, for example. Most are somewhere in between. The challenge is to demonstrate how models that we are accustomed to applying in one context can be useful in quite different contexts.

Necessity of mathematics: If models have a mathematic basis, they are more predictive (if they are validated and appropriate, data are available). But they have the disadvantage that the data to which they are applied must be quantitative. Qualitative models allow for more flexibility in making analogies and thinking about cause–effect relationships, but are not good for precise prediction.

Level of organization: Life and work exist at different levels or organization: the individual, the family or work team, the close community or company, the town/city, the state or nation, and the whole world. At the lower levels, individual people can be assumed to care for one another and share goals intimately; at intermediate levels, they share a common language and culture, while at the aggregate levels, they often do not even have common language, culture, or goals. Naturally, these differences must be taken into effect when models are devised.

Complexity: Simpler models (ones with few independent and dependent variables) are easier to understand, remember, and apply. The more complex the model (the more rules, the more variables in the objective function), the harder it is to apply and the less generalizable. It is common these days to refer to large societal systems as "system of systems," meaning that some subsystems (for which we have some sort of workable model) interact with other subsystems (for which we may have quite different workable models). Take, for example, a system of systems that is well short of society at large, an airplane. An aerodynamic model characterizes a resultant trajectory through the air as a function of speed and control surfaces. An engine determines thrust as a function of throttle setting, fuel combustion, and altitude. Mechanics determines how well the aircraft holds together as a function of materials and construction. Profit depends on ridership and costs, and so on. But there is no common model that encompasses all of "airplane," and the models within the different disciplines do not fit together well. How well a city functions is determined by separate models of terrain and layout, transportation and communication assets, profitable industry, an educated labor force, and fair and effective government. Two forms of models that are arbitrarily expandable (see Chapter 4) are networks (connecting simple dynamic transfer functions or probabilities) and fuzzy sets of rules (that can be aggregated to make simple inferences).

It is probably the case that the earlier described models that have been borrowed from engineering, such as those for feedback control, information processing, signal detection, and decision-making, have the greatest applicability to complex societal issues. They are underpinned by mathematics but also open to qualitative analogies. Concepts of probability and dynamics are surely relevant. Because of the many salient variables to contend with (and the difficulties of expressing experience in other than words), I would also assert that fuzzy logic should be included.

POTENTIAL OF FEEDBACK CONTROL MODELING OF LARGE-SCALE SOCIETAL SYSTEMS

Feedback control models are appropriate where the social system (controlled process in this case) has lags or time delays in responding to inputs. In this case, there is danger of overshoot and/or instability due to positive feedback that is out of synchrony with the controller effort to regulate and minimize error with respect to some target or ideal. This problem is certainly true of government regulation.

For example, if the government initiates increased spending to accelerate a weak economy, there is a long lag or time delay until the effects are fully realized, but over time the economy will pick up. By the time it is politically apparent that the economy needs no further boost and the government should slow down its spending, the economy will continue to accelerate from past spending and is likely to "overshoot" the desirable level. Eventually, the government will reverse course and curtail spending, and the same delayed effect will occur in the reverse direction. This is at least one of the reasons for economic fluctuation.

Governments seek to regulate energy production, pollution, military spending, military operations, and many other aspects of a functioning society, and there is always a dynamic lag in response, whether caused by public acceptance and compliance or other factors. Control theory calls for feedback not only of the level of the state variable but also its rate of change in order to achieve good response, especially to abrupt external influences. However, politicians and consumers are not well attuned to perceive and respond to rates of change.

THE STAMP MODEL FOR ASSESSING ERRORS IN LARGE-SCALE SYSTEMS

In Chapter 8, Section "Human Error," in conjunction with error and accident causation, there was mention of a particular model that had implications for large-scale socio-technical systems. The idea is that, whether in system

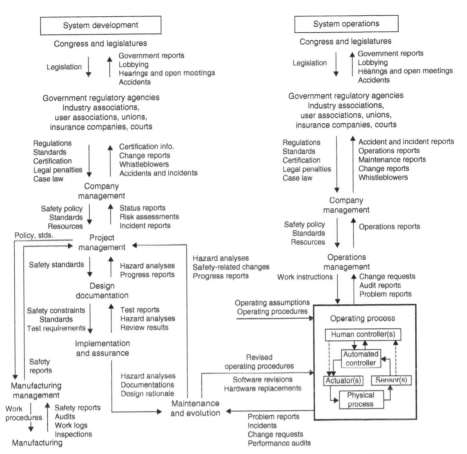

FIGURE 11.1 The Leveson STAMP model. Source: Courtesy of N.G. Leveson, personal communication.

development or system operations, there are many steps where things can go wrong, and where, if sufficient attention is not given to observation, reporting and remediation of small error precursors, these can lead to far greater problems downstream in the process. Figure 11.1 represents such a model, representing an active effort by Leveson (2012) to foster such thinking in the systems analysis and design world. The details of the figure speak for themselves.

PAST WORLD MODELING EFFORTS

Over the last 50 years, there have been many models purporting to predict the rise or fall of critical variables on a global scale. Perhaps the best known was a project funded by the Volkswagen Foundation and commissioned by the

Club of Rome, titled *Limits to Growth* (Meadows et al., 1972). It drew on work by a large number of participants who were experts in different areas, and modeled the interactions of five variables: population, industrialization, pollution, food production, and resources depletion as functions of time into the future. It employed a relatively simple approach based on *system dynamics*, originally attributed to Jay Forrester (Sterman, 2001). It was essentially a network of linear interactions incorporating positive and negative coefficients and first-order integrations (modern versions include probability, saturation, and other nonlinearities). The modelers considered several scenarios based on national policies and other externalities. One motivation was to show how simple linear extrapolations of independent variables lead to fallacious results because of interactions and resulting positive or negative effects of resulting feedback loops. Two scenarios predicted overshoot and collapse in the mid to late twenty-first century, while others predicted stability. The whole network was too complex to picture here. The project and its results were quite controversial at the time, with criticism of methods and assumptions as well as data used. There have been numerous updates by the original authors as well as others, and the predicted trends have roughly held true (see https://en.wikipedia.org/wiki/The_Limits_to_Growth). Figure 11.2 shows one example (coefficients, integrations, multipliers, and other mathematical operations not shown).

TOWARD BROADER PARTICIPATION

This book has played up the idea of models being "handles for thinking." We tend to ridicule purveyors of unique remedies for complex problems: "snake oil salesmen," "one-tune Charlies," or, in the present context, colleagues who "have model, will travel." Everyone would agree that having more metaphors, stories, analogies, or mental images in one's thinking arsenal is a desirable personal attribute. So also with scientific models of human–system interactions. But for these models to extend to large-scale societal issues they need to be refined, and more people need to become aware of their existence.

Of course, there already are academic, government, and other institutional researchers who work with models professionally and publish their findings in (often obscure) journals. Some small fraction of these studies are publicized in lay media. But the prevailing assumption has been that the "handles for thinking" that models provide is for the thinking of the professionals, and that most models are simply too sophisticated for the lay public to benefit from. I would challenge this assumption and assert that modern

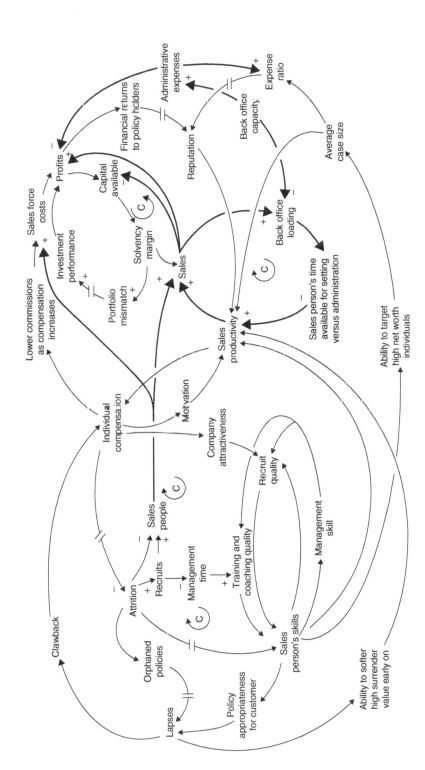

FIGURE 11.2 An example of system dynamics. Source: Used under https://commons.wikimedia.org/wiki/File:Causal_Loop_Diagram_of_a_Model.png.

communication and social media technology provide new opportunities to simplify and refine the models, and use them to educate and provoke thinking by a much broader audience. For example, the so-called MOOCs provide one such opportunity. MOOCS are now reaching thousands of users world-wide, some free ones being sponsored by university consortia (e.g., EdX), and some for nominal cost (e.g., Udacity). MOOCS can easily embody human–system models at whatever scale, but those for large-scale societal issues will surely be of interest to many more people than those for narrow cognitive engineering problems. Such models can provide *if-then-else* cause–effect demonstrations. The data input to such models can be prepro-grammed to make a point determined by the MOOC lesson-author, or can be driven by data input by the user.

The Internet is currently in process of making publicly available many sorts of data sets that are amenable for this function, provided the interface with the models is properly designed and human-factored. Further, the Internet and cell phone social media enable "crowd-sourcing" of model inputs in semi-real time, so that participants can have the sense of joint participation in modeling some social issue, for example, in meetings. Visualization of data sets is currently a rapidly developing technology (associated with the urgent need to assess "big data"). Rouse (2014) and Rouse et al. (2016) refer to "policy flight simulators" for exploring alternative policies and asking "What if?" questions, ranging from individual organiza-tions to national policy. The human–computer interface design is critical in making such simulators user-friendly. Figure 11.3 shows an overview of a policy flight simulator as defined by Rouse (2015). The diagram includes the multilayered feedback idea mentioned above, the point here being that the policy flight simulator should enable the participant to observe the systemic feed-forward and feedback interactions at each level.

There are other new technologies that can contribute to visualization of system phenomena in a dynamic (time-varying) sense. Computer graphic simulation enables visualization of trends and complex interactions to play out for an observer to watch, typically on an adjustable time scale (months or years sped up to seconds or minutes). VR technology allows the viewer to be "immersed" in virtual worlds that portray cause–effect relationships put in by the user (as contrasted to movies and TV shows that are preprogrammed). I can imagine VR exercises where a user experiences the comparative safety or efficiency effects of different policies regarding global warming or self-driving cars, or different designs for cities, and so on. Such VR exercises can be experienced as though one is alone in the virtual world or is simulta-neously together with avatars of family, friends, or other people who are online simultaneously, jointly asking the models "What if…?."

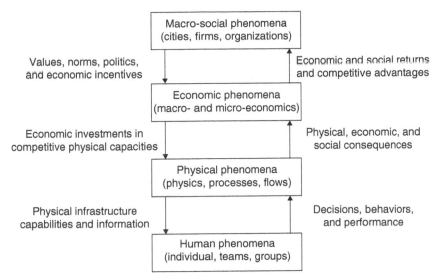

FIGURE 11.3 Relationships in a policy flight simulator. Source: Rouse et al. (2016). Adapted with permission of Elsevier.

The aggregate objective function (utility function), as discussed earlier in Chapter 6, Section "Valuation/Utility," is surely underlying such visualization and decision-making. Ever since the philosophers Jeremy Bentham and John Stuart Mill, the challenge of defining and optimizing human welfare has been implied, whether explicitly identified or not. Recently, the National Academy of Sciences has even undertaken to identify socio-technical metrics of "subjective well being," by which they mean to include "happiness, suffering, and other dimensions of experience" (Stone and Mackie, 2013). An ultimate challenge is to engineer a world that continually improves subjective well being. I think models can help.

APPENDIX

The following are explanations of the mathematics underlying the various models discussed before. Several sections discuss research issues without any mathematical treatment. Where math is presented, it is just enough to indicate the gist, and is certainly not a full and formal treatment, which would take many more pages than are allocated here. The words in many cases follow previous writings by the author of this book, some bearing the author's own copyright and some in previous Wiley publications.

MATHEMATICS OF FUZZY LOGIC (CHAPTER 4, SECTION "CRISP VERSUS FUZZY LOGIC")

In fuzzy logic, a descriptive word (i.e., tall or short or fast or clever) has a numerical degree of truthfulness or applicability as applied to some person or thing. The truthfulness or relevance number is called *membership*, and it ranges from 0 to 1 (see Figure A.1). In the context of a college basketball player, his 6′6″ height may be 0.8 for the fuzzy term *very tall* and 0.3 for the

Modeling Human–System Interaction: Philosophical and Methodological Considerations, with Examples,
First Edition. Thomas B. Sheridan.
© 2017 John Wiley & Sons, Inc. Published 2017 by John Wiley & Sons, Inc.

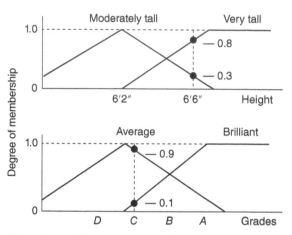

FIGURE A.1 Hypothetical fuzzy membership functions for basketball players.

term *moderately tall* on a hypothetical relevance scale. Grades may be a second consideration for his coach when selecting the team, who might want to distinguish brilliant players from those who are just average. The subject player might have mediocre C grades, and so have a 0.1 for *brilliant* and a 0.9 for *average*. Figure A.1 shows a plot of the hypothetical membership functions for the four fuzzy variables labeled, which can be applied to our subject basketball player.

So the coach may decide in his mind on a selection rule for who gets chosen for the team: "Select if a player is *moderately tall* AND *brilliant*, OR *very tall* AND *average* intelligence," the ANDs and ORs representing logical operations. For each candidate player and their fuzzy membership values, the rule is parsed and then evaluated, yielding a single relative rating number. The coach would then select the players with the highest ratings. The evaluation involves parsing the rule into component comparisons and then assigning the individual's membership numbers to the fuzzy terms in the rule.

Applying the rule and parsing, we get a player who is (0.3 AND 0.1) for the first part of the rule and (0.8 AND 0.9) for the second part. The convention is to take the minimum for AND and the maximum for OR. Thus, considering only what is in the parenthetical components in the first sentence of this paragraph, we get (0.1) OR (0.8) for the full rule in quotes given before. Then taking account of the OR between parentheses, we get a final score of 0.8 as a fuzzy ranking for this player. That would be compared to a similar scoring process for all other players. Then players with the highest scores would be selected.

MATHEMATICS OF STATISTICAL INFERENCE FROM EVIDENCE (CHAPTER 4, SECTION "SYMBOLIC STATEMENTS AND STATISTICAL INFERENCE")

The relative truth of any two hypotheses $H(A)$ and $H(B)$ can be derived and continually updated for new evidence D by using Bayes' theorem. This is done in conjunction with contingent probabilities. From basic combinatorics, we know that

$$p(A|B)p(B) = p(B|A)p(A) = p(A,B),$$

where A and B can be any two events. The vertical bar means "given what follows," so $p(A|B)$ means probability of A given that B is known to be true. $p(A,B)$ means the probability of A and B occurring jointly.

In its simplest form, Bayes' theorem is derived directly from the first two expressions, namely

$$p(A|B) = p(B|A)p(A)/p(B)$$

If we substitute H for A and D for B, where H represents "hypothesis" or "underlying truth" or "cause" and D represents "data observed" or "apparent symptoms" or "effect," we have

$$p(H|D) = p(D|H)p(H)/p(D).$$

This means that once some data D (evidence) is available for a case of known H, we can refine any prior estimate of probability of H, $p(H)$. This is done by multiplying $p(H)$ by $p(D|H)$ and dividing by $p(D)$ (the given Bayes' equation). $P(D|H)$ is what we know from a prior experience, that is, where we knew H to be true and then we observed D. $p(D)$ is the probability of that data being observed in general, independent of other circumstances.

Now suppose we observe some specific event D_1 and use the given equation to refine our prior knowledge $p(H)$ to get

$$p(H|D_1) = \left[p(D_1|H)p(H)/p(D_1) \right].$$

Then we observe D_2. It is now possible to use this equation again to determine the new expectation of H, $p(H \mid D_1 \text{ and } D_2)$, by substituting $p(H \mid D_1)$ as the current prior knowledge of H. That is,

$$p(H|D_1 \text{ and } D_2) = p(D_2|H)\left[p(D_1|H)p(H)/p(D_1) \right]/p(D_2),$$

where the term in square brackets is $p(H \mid D_1)$.

Suppose from the beginning we had two independent hypotheses, H_1 and H_2 and went through the given process. Then we could take a ratio of the two equations and would find that $p(D_1)$ and $p(D_2)$ would drop out since they are common to both equations, leaving

$$\frac{p(H_1|D_1D_2)}{p(H_2|D_1D_2)} = \frac{p(D_2|H_1)}{p(D_2|H_2)} \cdot \frac{p(D_1|H_1)}{p(D_1|H_2)} \cdot \frac{p(H_1)}{p(H_2)}.$$

The first term is called the "posterior odds ratio." The last term is called the "prior odds ratio." The two terms in between are called "likelihood ratios" for D_2 and D_1, respectively. This process can be extended to any number of likelihood ratios (they don't have to be independent of one another) to find a posterior odds ratio. In general, the posterior odds ratio is the product of all available likelihood ratios for all salient data, times the prior odds ratio.

Obviously, this formulation does not care about the order in which the data (evidence) are observed. It assumes that the process statistics are stationary (don't change with time). Old data count just as much as new data. If desirable, since the world is not always stationary, a Bayesian updating process can be adjusted to discount data the older it is.

MATHEMATICS OF INFORMATION COMMUNICATION (CHAPTER 5, SECTION "INFORMATION COMMUNICATION")

Shannon's information metric is simply the logarithm base 2 of the inverse probability $(1/p)$, assuming that after the message is sent its expectation probability is 1 (log base 2 of one is zero, i.e., no uncertainty). Thus, if one of two equally likely messages is sent and received (a 1-bit reduction of uncertainty), there is less surprise than if 1 of 16 equally likely messages is transmitted. (The latter is a 4-bit reduction of uncertainty, where the number of transmitted bits is $\log_2(16) = 4$ in this case.) Think of the game "20 questions," where the answer can be only yes or no. Assuming the questioner is good at framing questions so as to divide the possibilities into two equally likely categories, the question might be "Is the answer (one of the two categories)?" By successively narrowing categories, the answer can always be obtained. The number of yes–no questions is the equivalent number of bits of information.

Actually, transmitted information is but one of the useful measures of a message communication process. Information communication can be somewhat more complex, where information (variety) is lost or added in the process. Thus, one can distinguish between input information (uncertainly before the

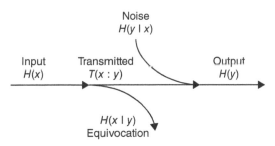

FIGURE A.2 Information relationships.

message is sent), equivocation (the potential uncertainty reduction that is never transmitted, i.e., variety that is lost in process), transmitted information (intended variety that would be reduced were it not for noise), the noise itself (uncertainly or variety added in the sending process), and output information (the uncertainty that would still need to be brought to zero after the message is received, because of corruption along the way due to equivocation or noise). These relations are shown in Figure A.2, where transmitted information equals input minus equivocation, or output information minus noise, the latter meaning output equals transmitted information plus noise.

The following are the key definitions concerning the information communication process diagrammed in Figure A.2, where subscript i refers to various inputs and subscript j refers to various outputs. Note that the equations below account for alternative messages that are not equally likely.

Input information:

$$H(x) = \Sigma \text{over } i \text{ of } p(x_i) \log\left[1 / p(x_i)\right].$$

Output information:

$$H(y) = \Sigma \text{over } j \text{ of } p(y_j) \log\left[1 / p(y_j)\right].$$

Noise:

$$H(y \mid x) = \Sigma \text{over } i \text{ and } j \text{ of } p(x_i, y_j) \log\left[1 / p(y_j \mid x_i)\right].$$

Equivocation:

$$H(x \mid y) = \Sigma \text{ over } i \text{ and } j \text{ of } p(x_i, y_j) \log\left[1 / p(x_i \mid y_j)\right]$$

Transmitted information:

$$H(x : y) = \Sigma \text{over } i \text{ and } j \text{ of } p(x_i, y_j) \log\left[p(x_i \mid y_j) / p(x_i)\right].$$

Information transmitted = input − equivocation = output − noise:

$$H(x:y) = H(x) - H(x \mid y) = H(y) - H(y \mid x).$$

A more extensive treatment of the problem, along with discussion of how this can be applied to human behavior, is found in Sheridan and Ferrell (1974).

MATHEMATICS OF INFORMATION VALUE (CHAPTER 5, SECTION "INFORMATION VALUE")

Information value (Howard, 1966) is different from information as defined by Shannon. Information value relates to the expected net value of investing in research (before knowing what the research will show about future events). It is assumed that the probabilities of states of the world are known, as are the payoffs that depend on both the states of the world and the actions taken. Viewed in another way, it is the value of clairvoyance: "if I knew the future now I would take a best action based on the state of the world. So what is it worth to me now to do the research that will tell me the future, though right now it is not known how the research will come out?"

Consider a simple numerical example: Let the future be X with probability 0.5, and Y also with probability 0.5. Assume that we know that if X occurs the payoff for action A is 4 and for action B is 7, and that if Y occurs the payoff for action A is 6 and for action B is 3. The problem is which of A or B to select to maximize payoff without knowing which of X or Y will occur?

If some research would tell me which of X or Y will occur one would select action B for X (yielding 7) and action A for Y (yielding 6). But knowing which X or Y will occur is not known until after the research is done. We only know that the probability if each is 0.5. So the best estimate of yield after the research is an expected payoff $(0.5 \times 7) + (0.5 \times 6) = 6.5$.

If one had no research option and would have to select A or B ahead of knowing which of X or Y will occur, one can only expect payoff for selecting $A = (0.5 \times 4) + (0.5 \times 6) = 5.0$, and payoff for selecting $B = (0.5 \times 7) + (0.5 \times 3) = 5.0$ also. So, in either case the best one could do is to get a payoff of 5.0.

Thus, even before knowing whether X or Y will occur, we know that it is worth $6.5 - 5.0 = 1.5$ to do the research that will let us take the best action depending on whether X or Y turn out to be the case. So, 1.5 is the *information value*. The cost of the research itself would have to be subtracted from 1.5.

The following generalizes the information value problem: Assume future states x have known probabilities of occurrence. Let $V(u_j \mid x_i)$ be the gain or reward for taking action u_j when a future x is in state i, where V can be money or time or energy or any measure of value. Ideally, a decision maker would adjust u_j (select j) to maximize V for whatever x_i occurs, yielding max on j of $V\left(u_j \mid x_i\right)$. In this case, the average reward over a set of x_i can be predicted to be

$$V_{avg} = \Sigma \text{over } i \text{ of } p\left(x_i\right)\left\{\max \text{ on } j \text{ of } V\left(u_j \mid x_i\right)\right\}$$

if the decision maker knows ahead of time which x will occur.

If the decision maker must select u_j ahead of time with no opportunity to select the ideal for whatever x_i that may occur, knowing x only as a probability density, $p(x_i)$, then the best a rational decision maker can do is to select u_j (ahead of time) to be the greatest expected value in consideration of the whole density function $p(x_i)$. In this case, the average reward over a set of x_i is

$$V'_{avg} = \max \text{ over } j \text{ of } \left\{\Sigma \text{over } i \text{ of } p\left(x_i\right) V\left(u_j \mid x_i\right)\right\}.$$

Information value, then, is the difference between the gain in taking the best action given whatever specific x_i occurs (and contingent of the probabilities of x_i), and the gain in taking the best action in ignorance of what x_i will occur. This difference is

$$V^*_{avg} = V_{avg} - V'_{avg}.$$

Almost always there is a cost to do the necessary research to know the future state X or Y. That cost should be subtracted from V^* to yield the final information value.

Note that information value is a "before the fact" valuation on doing the research. Naturally, if one knew the future state without research, one would simply act based on exactly what that state will be. But before the fact of doing the research, one is still dependent on the probabilities of what the research will show.

MATHEMATICS OF THE BRUNSWIK/KIRLIK PERCEPTION MODEL (CHAPTER 5, SECTION "PERCEPTION PROCESS")

Consider Figure 5.2. Mathematically, the correlation between judgment and state is

$$R = R_1 G R_2 + C\left(1 - R_1\right)^{0.5}\left(1 - R_2\right)^{0.5},$$

where R_1 is the state-to-observables correlation at left, R_2 is the perceived cues-to-judgment correlation at right, and G is the correlation between the cue weighting generated by the environment and the cue weighting corresponding to the judgment (Cooksey, 1996). The first term is the linear additive correlation from state to judgment. The second term is a correction factor, where C is found as the positive or negative correlation between the residuals of the human and environment models. As another correlation C is in the range $(-1, +1)$. Usually, C is found to be essentially zero, indicating that linear-additive models are sufficient to describe judgment. The advantage of this model is that it combines characterization of the environment with characterization of the human (Bisantz et al., 2000).

Measures other than correlation may also be utilized. For example, given the conditional probabilities of various cues for various states, Bayesian updating (see Appendix, Section "Mathematics of Statistical Inference From Evidence") can be employed to assess the relative probabilities of alternative states. Given joint probabilities for input–output combinations for each of the three cascaded stages, Shannon information theory (see Appendix, Section "Mathematics of Information Communication") can be used to analyze not only transmitted information but also information loss or equivocation (e.g., the shaded cue at left without counterpart cue at right) and noise (e.g., the cue at right without counterpart shaded cue at left).

MATHEMATICS OF HOW OFTEN TO SAMPLE (CHAPTER 5, SECTION "VISUAL SAMPLING")

Sheridan (1970a) sought a way of considering information content together with costs, where the supervisory operator observes a system state x and tries to take action u at intervals to maximize some given objective function $V(x,u)$. Specifically, the question is how often to sample x and/or update u. We assume that the state x in this case includes any directly observable inputs that will help the operator decide what further sample or action to take (a more inclusive definition than is usually assumed in control problems). The operator also is assumed to have statistical expectations about such signals.

The analysis is easier when x and u are scalars. Given assumptions about the probability of x and how rapidly it is likely to change, given a value (objective) function for consequences resulting from a particular x and a particular u in combination, and given a discrete cost of sampling, one can derive an optimal sampling strategy (see further) to maximize expected gain.

Assume that x has a known prior probability density $p(x)$, that x_o is its particular value at the time of sampling, and that $p(x_t \mid x_o, t)$ is a best available model of expectation of x at time t following a sample having value x_o. (Necessarily $p(x_t \mid x_o, t) = 1$ for $x_t = x_o$.) Thereafter the density $p(x_t \mid x_o, t)$ spreads out, approaching $p(x)$ as t becomes large, as shown in Figure A.3. Assume also that $V(x,u)$ is the reward for taking action u when the state is x. The goal is to maximize EV, the expected value of V.

Clearly, without any sampling, the best one can do is adjust u once and for all to maximize EV, that is,

$$\text{EV for no sampling} = \max \text{ over } u \text{ of } \left[\Sigma\text{over } x \text{ of } (V \mid x,u) \cdot p(x) \right].$$

If one could afford to sample continuously, the best strategy would be to continuously adjust u to maximize over u for each $(V\mid x,u)$ for the particular x encountered, so that

$$\text{EV for continuous sampling} = \Sigma\text{over } x_o \text{ of } \left[\max \text{ over } u \text{ of } (V\mid x_o,u) \cdot p(x_o) \right],$$

where $p(x_o) = p(x)$.

For the intermediate case of intermittent sampling, $E(V \mid x_o,t)$ at t after sample $x_o = \max$ over u of $[\Sigma\text{over } x_t \text{ of } (V \mid x_t,u) \cdot p(x_t \mid x_o,t)]$.

Then $E(V \mid t) = \Sigma\text{over } x_o \text{ of } E(V \mid x_o,t) \cdot p(x_o)$, remembering again that $p(x_o) = p(x)$.

In this case, for any sampling interval T and sampling cost C, the net EV is

$$\text{EV}^* = (1/T)\Sigma\text{over } t \text{ from } 0 \text{ to } T \text{ of } \left[E(V \mid t) - C/T \right].$$

So the best one can do is to maximize EV* with respect to T. These implications are shown in Figure A.3.

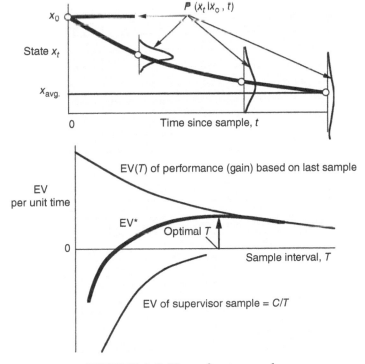

FIGURE A.3 How often to sample.

Sheridan and Rouse (1971) found in an experiment that even after the $V(x,u)$ functions were made quite evident to subjects, their choices of T were suboptimal relative to this model. Moray (1986), however, points out that subjects with even moderate training are quite likely to demonstrate suboptimal behavior, and only after they "live the experience of the costs" for a long time (as Senders' subjects did) are they likely to converge on optimality. Moray and Inagaki (2014) have provided an analysis with many similarities and some differences as compared to the one presented here.

MATHEMATICS OF SIGNAL DETECTION (CHAPTER 5, SECTION "SIGNAL DETECTION")

Consider a "truth table" or payoff matrix such as that shown in Figure A.4. This depicts a situation where the truth can take two forms, a signal mixed with noise (SN) or noise alone (N). The decision maker must respond with a best guess as to what the situation was. As shown, there are rewards (R) and costs (C) depending on whether the judgment was correct or incorrect. There can be prior subjective expectations (probabilities) of SN or N occurring.

The model assumes that the decision maker seeks to maximize the rewards minus costs, in light of what signal evidence e is experienced. The normative signal detection model assumes that the observer's brain conjures up two bell curves (Gaussian probability density functions) as shown in Figure A.5. The two hypothetical mental model curves are assumed to be displaced from one another by distance d' along an evidence axis e. The curves (vertical axis) indicate the probability (strength of evidence) for SN or N given a particular level of evidence e, where greater e monotonically favors SN over N. The distance d' is a function of the signal-to-noise ratio. TN means true negative and TP means true positive; MS means miss and FA means false alarm. So, according to the model, in consideration of both the perception and the

	Truth	
	SN	N
Decision — SN	R_{TP} True positive	C_{FA} False alarm
Decision — N	C_{MS} Miss	R_{TN} True negative

FIGURE A.4 Payoff matrix for signal detection.

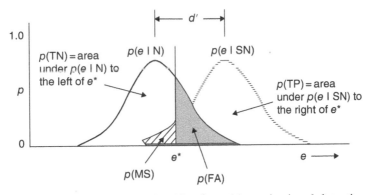

FIGURE A.5 Probability densities for evidence in signal detection.

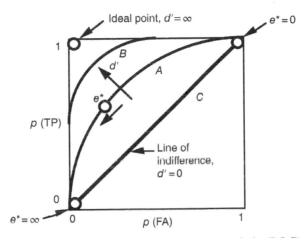

FIGURE A.6 Receiver operating characteristic (ROC).

relative rewards and costs, the decision maker will decide SN if the perceived e is greater than a threshold e^* and decide N if e is less than e^*. Clearly, if the reward for a true positive were very large and all other rewards and costs very small the decision maker would have to move the criterion e^* leftward, toward saying SN most of the time (fewer misses, but more false alarms). The following math specifies how the balance between prior probabilities, strength of evidence e in signal perception, rewards and costs is ideally made.

Another way of presenting Figure A.5 is by what is called a *receiver operating characteristic* (ROC) curve (Figure A.6). It is a cross plot of probability of true positive on the Y-axis and that of false alarm on the X-axis. These are the only independent variables, since $p(\text{TP})$ is $1 - p(\text{MS})$ and $p(\text{FA}) = 1 - p(\text{TN})$. Normative mathematical analysis states that an ideal decision maker will operate with a threshold criterion at some point (e^*) on one of the concave downward ROC curves shown in Figure A.6. Ideally, one would like to operate at the upper left corner, always ($p = 1$) deciding SN when the signal was a true SN (true positive) and never ($p = 0$) deciding SN when N is true (false alarm). Unfortunately, that is not possible because of the overlap in the perception curves in Figure A.5. Operating anywhere along the diagonal straight line in Figure A.6 indicates indifference (no signal discrimination ability, that is, the two bell curves of Figure A.5 completely overlap. The ROC curves move toward the upper left as the signal-to-noise ratio (d') increases. So the decision maker is stuck with operating along some one ROC curve. Where the decision maker puts the criterion e^* along the ROC curve is a function of the trade-off among the rewards and costs in consideration of prior probabilities. Keep in mind that Figures A.5 and A.6 are for idealized Gaussian functions. Actual experimental data would make for rougher functions.

Here is a mathematical derivation of the decision criterion e^*: Let $p(SN)$ be the prior probability of an SN and $p(N)$ be the prior probability of N. $p(N) = 1 - p(SN)$ since the two events are mutually exclusive and collectively exhaustive. To maximize expected reward one should decide SN when the expected value of doing that is greater than the expected value of deciding N. We use the underlined $\underline{SN}$ and $\underline{N}$ to represent the human decision and non-underlined SN or N to represent the actual condition, the truth. The expected value criterion is met when the expected value (EV) of deciding $\underline{SN}$ exceeds the expected value of deciding $\underline{N}$, namely $EV(\underline{SN}) > EV(\underline{N})$, or when

$$p(SN) R_{TP} - p(N) C_{FA} > p(N) R_{TN} - p(SN) C_{MS},$$

or when

$$p(SN)(R_{TP} + C_{MS}) > p(N)(R_{TN} + C_{FA}).$$

In a more useful form, decide $\underline{SN}$ when

$$\frac{p(SN)}{p(N)} > \frac{(R_{TN} + C_{FA})}{(R_{TP} + C_{MS})}.$$

Now remember that $p(SN)$ and $p(N)$ are prior probabilities before one gets additional evidence e from whatever source. After that bit of new evidence, we have better estimates, $p(SN|e)$ and $p(N|e)$. Thus the after-evidence criterion to optimally decide $\underline{SN}$ is when

$$\frac{p(SN|e)}{p(N|e)} > \frac{(R_{TN} + C_{FA})}{(R_{TP} + C_{MS})}.$$

Since $p(\text{SN}\,|\,e)\,p(e) = p(e\,|\,\text{SN})\,p(\text{SN})$ and $p(\text{N}\,|\,e)\,p(e) = p(e\,|\,\text{N})\,p(\text{N})$ according to Bayes' rule. When the two expressions are set in a ratio the $p(e)$ term drops out, and we have

$$\frac{p(\text{SN}|e)}{p(\text{N}|e)} = \frac{p(e|\text{SN})}{p(e|\text{N})}\frac{p(\text{SN})}{p(\text{N})}.$$

With substitution and rearrangement of terms, and with only $p(e\,|\,\text{SN})\,/\,p(e\,|\,\text{N})$ on the left side of the inequality, we get the decision criterion in a most useful form: decide <u>SN</u> if

$$\frac{p(e|\text{SN})}{p(e|\text{N})} > \frac{p(\text{N})}{p(\text{SN})}\frac{\left(R_{\text{TN}} + C_{\text{FA}}\right)}{\left(R_{\text{TP}} + C_{\text{MS}}\right)}.$$

The left side of this equation is the so-called *likelihood ratio* for any evidence *e*, where the farther one moves to the right along the *e*-axis in Figure A.5 the larger is that ratio. It is important to mention that the only requirement on *e* is that it be scaled monotonically with the likelihood ratio. The right side of the equation is a constant generally called Beta in signal detection theory. More on this topic is found in Green and Swets (1966).

RESEARCH QUESTIONS CONCERNING MENTAL WORKLOAD (CHAPTER 5, SECTION "MENTAL WORKLOAD")

There has been a continuing interest in mental workload, both because of its practical importance in tasks such as aircraft piloting and air traffic control, car driving and driver distraction (and the potential for self-driving cars), and other tasks where the pace of activity is set by factors not under the operator's control. Moray (1979) provides a number of views on mental workload, including both its definition and its measurement. Hart and Sheridan (1984) review the definition and measurement of workload as regards automation. They point out that automation has been motivated in some measure by desire to reduce workload, in some cases by people oblivious to the fact that the requirement for human monitoring of the automation itself creates workload (hence the pilot's term "killing us with kindness"). Mental workload is seen as a function of many factors, some of which have nothing to do with the operator: task objectives, temporal organization and demands (pacing, urgency, ability to select subtasks), hardware and software resources available (including automation), and environmental variables (workstation geometry, ambient temperature, vibration, lighting, etc.). Other factors are operator dependent: operator perception of task demands (which may be quite different from what is nominally assigned), operator qualifications and capacities, operator motivation (both *a priori* and in terms of accessing feedback during the task), and actual operator behavior.

Hancock and Desmond (2001) provide an extensive review of factors covering not only mental workload but also closely related issues of stress and fatigue. Hancock and Chignell (1988) plot mental workload in three dimensions, whose axes represent effective time for action, perceived distance from desired goal state, and level of effort required to achieve the time-constrained goal. This representation allows them to specify workload contours that incorporate the factors of operator skill and effort.

Experimenters have tried to understand how workload builds up as operators undertake multiple tasks, and how workload is affected by manual, perceptual, and mental components of behavior. Moray (1979) provides an excellent review of the earlier literature on mental workload.

Common sense dictates that if a human operator is given too little to do, then the human will become inattentive and the monitoring performance will decline. The classical literature on vigilance generally supports this claim (Buckner and McGrath, 1963). If there is too much workload, as was discussed before, performance also declines. Thus there must be an optimum somewhere between the extremes of task load. The so-called inverted *U*-curve hypothesis (discussed earlier in Chapter 5) is not a bold hypothesis at all. However, finding just where such an optimum exists for any particular task has been an elusive goal, and many factors other than task load seem to be at work. Verplank (1977) explored this question in the context of manual control and found little support for a specific Yerkes and Dodson (1908) "inverted-*U* hypothesis" in any simple form.

Wierwille et al. (1985) reported a series of evaluations of the sensitivity and intrusion of mental workload estimation techniques. In simulated aircraft piloting, they imposed psychomotor tasks (wind disturbances to be nulled), perceptual tasks (detection of instrument failures), cognitive tasks (wind triangle navigation problems), and communication tasks (execution of commands, response to queries). They measured subjective workload ratings, several attributes of primary task performance, several aspects of secondary task performance, and various physiological indices. They found that subjective ratings were highly sensitive to changes in imposed workload. Primary task performance measures (e.g., aircraft response) were not sensitive, except for those measures that corresponded to the operator's control movements. Time from presentation to response was sensitive for the perception and cognitive tasks, even better than the subjective ratings (complex demands simply took longer to fulfill). Conversely, communication response times were shortened for increased task load. Most secondary task measures fared poorly (e.g., mental arithmetic, memory scanning, regularity of tapping), except for the standard deviation of time estimation. Secondary task measures cannot be used simultaneously with subjective ratings, as the latter are affected by both primary and secondary tasks. Physiological indices (pulse rate, respiration rate, pupil diameter, eye blinks, and eye fixations) were mostly insensitive to imposed taskload.

Wickens et al. (1985) did extensive workload-related experiments in one-axis and two-axis tracking (two hands in the latter case). The instructions were to pay special attention to the designated primary axis, but never ignore the secondary axis. Scores based on simple trade-off functions were communicated to subjects after each trial. The investigators found that subjects' allocation of attention was distinctly nonoptimal. Apparently, as tracking on the secondary axis became more difficult (the dynamic order increased), resources were allocated to it disproportionately to what was optimal by the scoring function. The investigators comment that their subjects had a hard time reallocating their efforts among concurrent tasks of different and changing difficulty. Also, there seemed to be an added "overhead" cost of time-sharing among multiple tasks.

Berg and Sheridan (1985) performed a piloting simulation experiment with different approach and landing scenarios which emphasized the effects of manual and mental activity separately and in combination. They measured altitude and speed deviations of the aircraft, and subjective ratings of "activity level," "task complexity," "task difficulty," "stress," and "mental workload." Their results indicated that relative to a baseline scenario of low manual and mental activity, the high-manual-workload scenario most affected subjective ratings, whereas the high-mental-workload scenario most affected aircraft performance. Subjects seemed more conscious of the effects of manual activity than of those of mental activity. The increases in workload ratings proved more different from each other as manual activity increased, less so as mental activity increased. Relative to the baseline, the increase in workload ratings for the high-manual-workload scenario and the increase in ratings for the high-mental-workload scenario were approximately additive for the combined manual and mental scenarios.

Ruffell-Smith (1979) studied pilot errors and fault detection under heavy cognitive load and found that crews made approximately one error every 5 min.

BEHAVIOR RESEARCH ISSUES IN VIRTUAL REALITY (CHAPTER 5, SECTION "EXPERIENCING WHAT IS VIRTUAL; NEW DEMANDS FOR MODELING")

As suggested in Chapter 5, Section "Experiencing What Is Virtual; New Demands for Modeling," the existence of virtual reality (VR) technology enables research that impinges on the question of what we perceive to be reality. These days we are bombarded by computer-generated TV ads and films, doctored still photographs, and other means to produce images that we know are not real. It becomes more and more difficult to determine what is real.

In some primitive forms, the art of eliciting a VR experience has been around for a very long time. Verbal story-telling probably dates back to primitive peoples who lived in caves. One can easily imagine that good story-telling, whether to groups huddled around a campfire or to children by parents at bedtime, has always stirred the listeners and conjured up active and realistic mental images. After the invention of writing, there were depictions of battles and other events that actively provoked the reader's imagination. Later came the theater. Much more recently came radio, and who of us oldsters can forget radio serials such as the Lone Ranger, the Shadow, and other serial radio programs that kept listeners glued to the set, filling in for visual images that were not there. All of these communication media can be said to have elicited a VR experience.

One can conclude from the active development and ready market for today's VR technology that there is a strong interest by people to engage in VR experiences, perhaps to get away from the somber realities of the real world. VR is enjoyable and memorable. Also we note two attributes of human behavior that have been shown to enhance the VR experience. First, active bodily participation (e.g., looking around) helps make the virtual seem real. Second, we note that active voluntary suppression of disbelief (that the VR is not real) has a strong effect on enabling psychological "immersion" in the experience.

Philosophical tradition offers contrasting perspectives regarding what is real. One is what is commonly called Cartesian or mind–body dualism: that the brain processes perceptions of the outside world and these are passed to an immaterial "mind." The latter is attributed to French philosopher Rene Descartes because he postulated himself as a *res cogitans*, usually translated as a "thinking substance": he was conscious and existed because he could doubt. However, there is controversy as to what he meant was located in the "mind."

Modern science mostly assumes that there is a true material reality "out there" and by means of the scientific method we struggle to bring our scientific and our mental models to be ever closer approximations to the external truths. The brain is surely material and exists. Most scientists do not consider the "mind" to exist apart from the functioning of the brain. There are objective measures of the external environment and subjective measures of what we are thinking.

The German philosopher Martin Heidegger rejected the Cartesian view and asserted that all meaning, hence all reality, is conditioned by interpretation, including the beliefs, language, and practices of the interpreter. According to Heidegger, we are "thrown" into situations where action is unavoidable (*throwness* in Heidegger terminology), the result of such action is unpredictable, and stable representation of the situation is not possible. In normal use of a tool or other object (e.g., in hammering), the tool becomes transparent to the user, who then cannot conceive of the tool independently (it *is ready-to-hand* in Heidegger-speak). However, if some abnormality occurs (e.g., the hammer slips) there *is breakdown*, and the tool can then be conceived in the "mind" (it becomes *present-at-hand*). Normal "being," in Heidegger's view, means complete involvement in a dynamic interaction in which subject and object are inseparable. Only by stepping back and disconnecting from the involvement can a person perceive the separate elements of the situation.

Seemingly related to the Heidegger view are ideas put forth by American psychologist J.J. Gibson (1979). According to Gibson, perception is the acquisition of information that supports action, especially with regard to overcoming constraints on action. Gibson calls this constraint-conformance *affordance*. Actions affect the environment, and the environment in turn affects the action in complete reciprocity. Perceptions are true, as Gibson sees it, to the extent that they support action in the environment.

I find the Heidegger and Gibson views compatible with one another. The Heidegger–Gibson perspective is credible in the sense that all perception is based on the result of previous action and learning (with exception of instinctual perception and response). At the same time, there is a credible view that there must be some reality "out there" that forever must be unknown in many (most) details—no chance to unravel it no matter how much science is brought to bear for any finite time into the future.

The implications of these issues for cognitive engineering are unclear at this point but seem bound to become more salient as the VR technology advances.

MATHEMATICS OF HUMAN JUDGMENT OF UTILITY
(CHAPTER 6, SECTION "VALUATION/UTILITY")

To understand the accepted means of defining and measuring utility (relative worth) start with utility in its simplest form. The Von Neumann axiom specifies that the utility U of an object or event C has meaning relative to a lottery of two other objects or events and their known utilities. Define

$$U(C) = pU(A) + (1-p)U(B),$$

where utility U is measured in some abstract dimension, sometimes called *utiles*. We shall see that this equation has its meaning only in relative terms, where p and $(1-p)$ are probabilities of mutually exclusive and collectively exhaustive events A and B that bracket C on some attribute (height, weight, cows, or potatoes, etc.)

As an example assume that A and B are values in dollars, and let A and B be extremes of the range of the dollars attribute on which we wish to scale utility, say $A = \$100\,000$ and $B = \$0$. Define A to have utility 1 and B to have utility 0. If we were to let $p = 0.5$, then, according to the definition of utility, $U(C) = 0.5$. In terms of Von Neumann's axiom, that means we would be indifferent between having C for sure and a 50–50 lottery (chance) of \$0 and \$100\,000, as represented by the upper part of Figure A.7.

The question is as follows: where does C lie on the dollar scale? Implementing an indifference judgment experiment, most of us would select a point along the horizontal physical metric axis in Figure A.7 significantly

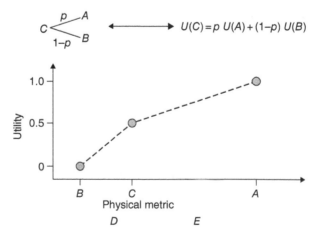

FIGURE A.7 The definition and experimental elicitation of a person's utility function.

less than $50 000, as shown. We would not want to risk gaining nothing, assuming we are only given this one opportunity. From this, we note that the utility curve would not likely be a straight line.

Such a concave downward curve means the decision maker is risk averse, opting for something less than expected value in order not to risk gaining nothing. Risk aversion is why we purchase insurance, paying the insurance company a relatively small amount to avoid the risk of a very large consequence. The insurance company, by insuring many clients, can afford to play on an expected value basis. The individual risk-taker cannot afford that luxury.

This same procedure as before can then be repeated to determine other points on the utility curve. To determine the location of a point D that represents a utility of 0.25 between B and C (Figure A.7) the definition and experimental elicitation of a person's utility indifference judgment experiment is repeated. Now C and U(C) are taken to represent the maximum of the range on each axis, respectively, with B and U(B) the zero point, and a subjective judgment is made of where D would lie for the judge to be indifferent to a 50–50 lottery of B and C dollars. To determine the location of a point E between C and A representing U = 0.75, C and U(C) are taken to represent the minimum of the range, with A and U(A) at the maximum, and another 50–50 indifference judgment is made. Theoretically, one can continue in this manner, but because of experimental error only several points are usually sufficient between 0 and 1 for any attribute, such that a reasonable curve can be drawn.

It should be noted that the procedure as summarized here assumes smooth monotonicity between the anchor points initially assumed (B and A in this example). But consider for some attribute there is a known maximum utility in between the anchor points, say at F. This could occur, for example, if one is determining a utility function for desired height or weight or the size of a hamburger, where an intermediate value at F is most preferred to either extreme. In this case, two utility functions should be determined, one between B and F, the second between F and A.

MATHEMATICS OF DECISIONS UNDER CERTAINTY (CHAPTER 7, SECTION "DECISION UNDER CONDITION OF CERTAINTY")

Consider Figure A.8. This depicts what is called a *decision space*, in this case with points (letters) representing choice alternatives. For example, let us assume a decision maker is deciding between used trucks, any of which might do the job.

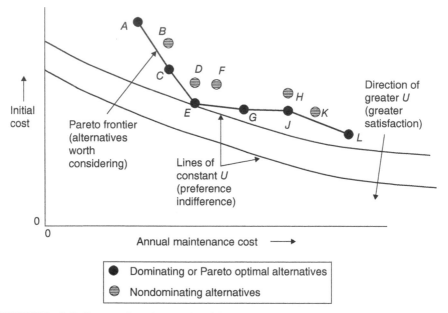

FIGURE A.8 Pareto frontier and utility curve intersection determine optimal choice.

To make the example simple, only two attributes are shown, maintenance cost and initial cost, assuming the decision maker is mostly concerned with cost. We assume that cost is defined in dollars. In general, there are more than two attributes of any decision, so the decision space is multidimensional, which cannot be represented here in a simple graph. Assume in this example that only a finite number of trucks is available, so the decision maker must decide among those combinations shown by the dots.

Note that some choices dominate other choices, that is, for the same maintenance cost truck *C* has lower initial cost than truck *B*. A rational decision maker would always prefer *C* over *B*. Truck *E* similarly dominates *D*. *J* dominates *H* for initial cost and also *K* based on maintenance cost, where initial cost is the same. *A*, *G*, and *L* are not dominated, assuming *G* has slightly higher initial cost than *J*. Together, the trucks depicted by solid dots comprise a set of nondominated alternatives which trade-off initial cost and maintenance cost. No purpose is served by considering the dominated alternatives, since the total cost for them is still greater than the others that are just as good in other respects.

The example assumes that the decision maker can get lower initial cost only at higher maintenance cost. The rational decision maker rejects whatever is up and to the right of the heavy line, and wants to be down and to the left. Unfortunately, there is no truck with lowest initial cost AND lowest maintenance cost. The points on the heavy line are said to form a *Pareto frontier*, named after an Italian economist Vilfredo Pareto (Barr, 2012). The Pareto frontier defines the best that can be had, the set of alternatives from which to choose, but still leaves the decision maker having to decide among the points on this line.

How to decide? The rational choice is to select the alternative with the greatest utility as defined in Appendix, Section "Mathematics of Human Judgment of Utility." Assume that the two light lines are hypothetical constant utility curves, determined by the Von Neumann procedure described above. Each is a line of utility indifference, defining a trade-off between the two cost attributes. For a set of such lines "better" is down and to the left, since the attributes are costs. Therefore, the best choice is the point of intersection of the Pareto curve with the indifference curve of best achievable utility, in this case point *E*. Of course, one need not draw the Pareto curve explicitly, since one could evaluate *U* of each truck on the Pareto curve and go right to the alternative with greatest utility. But when there are a large number of alternatives, or a continuous space, it is helpful for the decision maker to consider the Pareto frontier explicitly.

Note that there are two kinds of trade-off curves in decisions under certainty: (1) the Pareto frontier, which rejects dominated alternatives and represents a trade between alternatives based only on the direction of better or worse, but has nothing to do with the subjective utility trade between attributes; and (2) the utility indifference curves, which represent the subjective utility of dollars and the trade between attributes, but say nothing about available choice alternatives.

MATHEMATICS OF DECISIONS UNDER UNCERTAINTY (CHAPTER 7, SECTION "DECISION UNDER CONDITION OF UNCERTAINTY")

Decision under uncertainty means that the payoffs for different decisions occur with probabilities. This situation is represented in Figure A.9, a *payoff matrix*. In this example, there are three decision options shown, *A*, *B*, and *C*, and two independent contingencies (not under the decision maker's control), *X* which is known to occur with probability 0.3, and *Y*, which is known to occur with probability 0.7. The numbers in the cells are the payoffs for each combination of contingency and decision. Payoffs can be stated in dollars, or in terms of other quantities such as time or effort, or in utility, as described earlier.

Contingencies

	X $p(x) = 0.7$	y $p(y) = 0.3$	Expected Value	Max	Min
A	5	5	5.0	5	5
B	7	0	4.9	7	0
C	6	3	5.1	6	3

(Decisions)

FIGURE A.9 Sample payoff matrix for decisions under probabilistic contingencies.

If one expects to make repetitive decisions in the same environment of contingencies and payoffs, one normally selects that decision that has the highest expected payoff (see first column at the right in Figure A.9). This would be decision C, where expected value is $(0.7 \times 6) + (0.3 \times 3) = 5.1$.

However, there are other criteria that may apply, as indicated by the max and min columns to the right. A risk-taker may be attracted to a selection based on what will yield the maximum possible payoff, and choose B hoping for a payoff of 7. More likely, and particularly if this is a unique decision not to be repeated, the decision maker may choose A, which is the best of the worsts, the most conservative decision, the one that minimizes the downside risk. This called a *minimax* decision.

Buying insurance is an example of an everyday minimax type of decision. The insurance buyer pays a small premium of money to avoid the risk of a great loss, even though on an expected value basis the insurance buyer would get a greater payoff. The insurance company, by repeating many insurance transactions and settlements, operates on expected value and makes money in doing so.

MATHEMATICS OF GAME MODELS (CHAPTER 7, SECTION "COMPETITIVE DECISIONS: GAME MODELS")

There are many kinds of formal games. Games can be two-player or multi-player. The examples here are only of the two-player type. Consider the two payoff matrices of Figure A.10. Assume that, instead of fixed probabilities and payoffs (as in decision under uncertainty), the payoffs are determined by actions of an intelligent opponent. The first number in each cell is the corresponding payoff (or penalty) to one's self and the second number is the

	Opponent decision	
	X	Y
A	−3, 3	−1, 1
B	−4, 4	−5, 5

Own decision (left matrix)

	Opponent decision	
	X	Y
A	−2, −2 / R, R	−10, −1 / S, T
B	−1, −10 / T, S	−6, −6 / P, P

Own decision (right matrix)

FIGURE A.10 Dominating and nondominating strategies (at left) and prisoner's dilemma (right).

payoff (or penalty) to the opponent. Each player is trying to maximize his payoff. Both examples used here are for zero-sum games, where one party's gain is the other party's loss. (This need not be the case; the payoff numbers could take any values.)

Typically, a *minimax* strategy (as defined in Appendix, Section "Mathematics of Decisions Under Uncertainty") is employed in zero-sum situations, choosing the option with the least downside risk, the most conservative strategy. For the matrix at left in the figure one's own minimax is *A*, since it is the best of the worst outcomes, depending on what the opponent does. (If one's own decision is *A* the worst of −3 and −1 is −3. If one's own decision is *B* the worst of −4 and −5 is −5, so the best of −3 and −5 is −3, pointing to *A* as the minimax.) The opponent's minimax is *X*; (the worst of 3 and 4 is 3 for *X*, the worst of 1 and 5 is 1 for *Y*, so the best of these is 3, pointing to *X*).

With these particular (arbitrarily chosen) numbers, *A* happens to be a dominating own strategy, since *A* is better than *B* for both *X* and *Y* moves of the opponent. The opponent, however, has no dominating strategy, since, while *X* is better for him when you choose *A*, *Y* is better for him if you choose *B*.

A game situation with an interesting history and moral lessons is called *prisoner's dilemma*. This game, at the right in Figure A.10, is named for the dilemma a prisoner has in deciding whether to keep silent about his fellow prisoner (cooperate) or provide harmful testimony about him and gain a lighter sentence (defect). In the matrix at right, the upper left cell is mutual cooperation, both keeping quiet about the crime (both rewarded *R* with a light sentence). The lower right cell is mutual defection, testifying against the other (both punished with *P*). The other two cells are where one testifies *T* against the other and wins a light sentence, while the other tries to cooperate but becomes a sucker *S* with a heavy sentence. Thus, if you defect (choose *B*) while your colleague tries to cooperate (chooses *X*) you lose little and your colleague suffers great loss. If you both cooperate (A, X) you both suffer only a moderate loss, while mutual defection (B, Y) results in moderate punishment. Note that both players have minimax and dominating strategies (B, Y): mutual defection. Yet paradoxically, this conservative minimax result (-6 for both players) is significantly worse than the more risky mutual cooperation result (-2 for both). A prisoner's dilemma situation is defined by the constraints $T > R > P > S$ and $R > 0.5(S + T)$.

This paradoxical situation is not unlike many situations in politics, business, and personal life where each party's effort to get the advantage over another party can result in greater cost to both parties than if they cooperate to mutual advantage. But the temptation to defect is so rational! There seems to be a great moral lesson in this simple game, which many researchers have studied and writers have commented upon.

MATHEMATICS OF CONTINUOUS FEEDBACK CONTROL (CHAPTER 8, SECTION "CONTINUOUS FEEDBACK CONTROL")

Referring to Figure 8.2, the following algebraic calculations will make some essential points about how feedback control works.

Each of the G terms in Figure 8.2 is an input–output functional relationship (transfer function) couched in differential equations. G_c represents the controller (e.g., driver or pilot or mechanical controller) and G_p represents the controlled process, the car or airplane.

$$x = G_p(u + w) = G_p G_c(r - x) + G_p w,$$

assuming y is a perfect measurement of x. Collecting terms and rearranging give the transfer functions for output x in terms of goal input r and disturbance w,

$$x = \left[(G_c G_p)/(1 + G_c G_p)\right]r + \left[(G_p)/(1 + G_c G_p)\right]w.$$

In a simple case, the controller G_c is an output/intput ratio or gain coefficient K and the controlled process G_p is an integrator (e.g., a fixed steering wheel input makes the car direction change at a fixed rate) represented algebraically (common Laplace transform notation) as $(1/s)$,

$$x = \left[(K/s)/(1 + K/s)\right]r + \left[(1/s)/(1 + K/s)\right]w,$$

which, when numerators and denominators are multiplied by s, yields

$$x = \left[K/(s + K)\right]r + \left[1/(s + K)\right]w$$

Note that as the controller gain K becomes large the closed-loop transfer function x/r approaches 1 (which means, e.g., the car follows the twists and turns of the road perfectly), while x/w approaches 0 (which means the disturbance is completely cancelled).

Thus, the simple solution to automation might seem to be negative feedback with as great a gain as possible, to be as sensitive as possible to any error between $r(t)$ and $x(t)$. Unfortunately, it is not quite so easy, as gain that is too great will lead to instability. That requires further considerations that we will not detail further here.

The reader is referred to wikipedia.org/wiki/control theory for a more extensive treatment of feedback control.

MATHEMATICS OF PREVIEW CONTROL (CHAPTER 8, SECTION "LOOKING AHEAD (PREVIEW CONTROL)")

One model for preview control is a *dynamic programming* algorithm for determining the best path within the range of the preview (and then iterating for a next segment). Such a model does not imply that such an algorithm is the cognitive mechanism used by people, only that it provides a normative basis against which to compare subject data.

How does dynamic programming work? Here is a simple example. Assume, at each of four stages in time after an initial stage, that there are three possible states of a system, represented in the figure by circles. From each state at a given stage the incremental cost to get to each state at the next stage is some function of the two states, represented in the example in Figure A.11 by a number on the line connecting the two states. Starting at

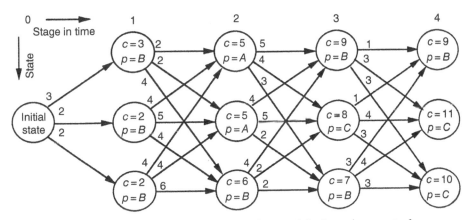

FIGURE A.11 Dynamic programming model of preview control.

stage 1, the cumulative cost (c) is indicated within each state (circle), this being the incremental cost to get from the previous state. At stage 2, the cumulative cost (c) indicated within each circle is the least cost of the three ways to get there, considering the cost to each of the previous three states plus the corresponding incremental cost along the connecting line. That least cost path is also indicated in each circle by the appropriate letter, A, B, or C, starting from the top. This simple procedure is repeated for the other stages. The result (working backward and noting which circle is best to have come from) is a least-cost trajectory that includes 4(top), 3(middle), 2(bottom), and 1(middle).

Note that the only information that needs to be saved at each incremental stage is what is in the circle, including cumulative cost and which previous circle is best to have come from. Further, for T stages and N states per stage, only TN comparisons need be made, far fewer than computing and evaluating all possible paths.

STEPPING THROUGH THE KALMAN FILTER SYSTEM (CHAPTER 8, SECTION "CONTROL BY CONTINUOUSLY UPDATING AN INTERNAL MODEL")

The following is a very qualitative description of how a Kalman filter or estimator works (Figure A.12). It is provided because this author feels that it has a lot to suggest about how human learning works. We start with the notion that in the external physical environment lies a true reality that that can never be known fully (the Physical reality blob on the right) of Figure A.12. It is important to understand that by a "reality" is meant the full set of physical cause–effect relationships in the slice of the world that is being considered, how any action results in a new state. This is more than simply a measurement (a single snapshot) of the state of the world in one time or place. By itself such a snapshot cannot tell the computer much of how the world "out there" really works. But by test actions on the world, one can see how it responds over time and with different inputs, and by enough such actions one can derive a good model of the world's action–response characteristics.

The true physical reality, of course, consists of an infinite amount of cause–effect information (potential-action-to-new-state transformations). The intention is to construct an "internal model" of some subset of that reality (a clear instantiation of model-dependent reality as mentioned in the Introduction). The internal model can at best represent only a minute fraction of the cause–effect information in the true reality, cast in terms of only a few dimensions of interest (i.e., variables of interest to the engineer designing the control system or to the person whose perception we design for). In the engineering version, the internal model of the environment is built into a computer, whereas in our diagram we suggest that the internal model can also be mental, in the head of the person confronting the considered slice of physical reality. The internal model is the key component within our larger model of a person's belief acquisition.

At the center left in Figure A.12 is a rectangular block in bold outline labeled "Internal (mental) model of reality." Thus at the heart of our overall model of perceiving reality lies another model (a model within a model). The internal (mental) model is where the above-mentioned small slice (of the full set of cause–effect relations) is represented, corresponding to the best information a person has momentarily on how the world works. The current state of the mental model (its output) is a momentary value of a small set of variables that specify the best estimate of reality (in the engineering version a mathematical vector). It is represented by an arrow from the bottom of the block. This triggers a policy decision, namely an intention of what action to take to modify the external environment in a desirable direction. In engineering, this "State-based policy deciding on action" is called a *control law*. It says if X is the case then take action Y.

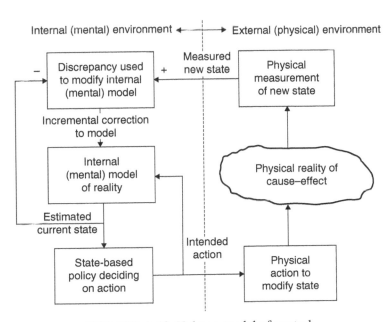

FIGURE A.12 Kalman model of control.

Following the arrows, the resulting intention signal is sent to the physical environment (through what corresponds to motor neurons) to the "Physical action to modify state" block (which in a person is muscle). Note that this same signal of intended change is sent back to the internal model, so that the internal model can be updated to a state corresponding to what state it thinks the physical reality will take when the intention is realized in the external world.

Thus, the (human or mechanical) muscle produces an actual change in the "Physical reality of cause–effect." This is usually a small incremental change, only enough to make a small change in the "Physical measurement of state" (performed by physical sensors in the engineered system or in human sensors).

Now the question is whether the internal (mental) model really has a true perception of what is going on with the physical reality (at least that very small segment of reality that the internal model is concerned with). It did update itself based on the intention resulting from the change policy—a kind of dead-reckoning. But this update was blind to what *really* might have happened when the intention was transformed by the muscle to make a real impact on the physical reality. To set the record straight another bit of ingenuity is involved, and that is to compare what actually happened (Measured new state) to what was thought would happen (Estimated current state). The two are compared by the block at the upper left labeled "Discrepancy used to modify internal (mental) model." Then that discrepancy result is used to tweak the model to better conform to what actually happened.

This is the raw outline of the estimation process. The process continues in small increments, thus enabling the internal (mental) model of reality to hone in, or to evolve through successive interactions, to an ever more valid representation of reality, at least in terms of the variables of concern. It combines open-loop *dead-reckoning* with feedback control. Realize that simple feedback control amounts to blind error-nulling without any effort to build an internal model of the external reality. In control engineering practice, the logic of what does or could go on inside the two blocks above and below the internal model block are somewhat complex and the subject of ongoing research. Further, because the physical action and physical measurement blocks above and below the physical reality block may be less than perfect (time delays, noise, etc.), there are additional elements that help, such as a computer model of the measurement process. But the above description should provide sufficient explanation for the reader to appreciate how learning can be modeled as an interactive "bootstrap" improvement process. In this process, perception and learning are clearly contingent on what actions are taken to modify the external reality, and what measurement is made on same. In control engineering, this estimation technique is also called an *observation* process.

The estimation model can be used further to reflect on what we humans do when we "behave." Suppose a person could sense the external environment perfectly and continuously in time. This would necessitate accurate sensing for vision, hearing, touch, and muscle position and velocity. But suppose that person was unable to relate what was sensed to any internal model of the environment, any map, any procedure. The person would be reduced to simple error-nulling feedback control.

Now suppose at another extreme our person had no sensors that look outside to see or measure the external environment, but the person had a perfect internal model of the environment at some initial point in time. Suppose also that there was perfect muscle control, so that the person could calculate what would result from any and all responses that were made. Thus, starting from an initial position, by perfect dead-reckoning our person could know the resulting location exactly no matter what action was taken, even though being blind to the actual environment. But if something in the outside world changed the internal model would no longer be valid. The person would have no way of knowing about new obstacles and thus any dead-reckoning calculations would not provide an accurate prediction of location.

Both the feedback control and the dead-reckoning paradigms outlined above assumed that either exteroceptors (that measure the outside world) or interoceptors (that measure and control the actions of the muscles or motors) were perfect. However in the estimation model of perception and learning, the sensory mechanisms can be imperfect and noisy and the system will still hone in on an accurate model of cause–effect relations. A real person makes many movements open loop for short periods of time, but intermittent feedback comes to the rescue and the internal model eventually provides a decent estimation of the external reality.

This paradigm of estimation clearly represents a means to continually cope with an external environment that may be slowly and continually changing. It is fully analogous to the process of science itself, in the sense that science must gradually chip away *through interaction* to refine any model of what is being investigated. The interaction is what enables the resulting gain in knowledge of the world and control within it.

REFERENCES

Anderson, J.R. (1990). *The Adaptive Character of Thought*. Hillsdale, NJ: Lawrence Erlbaum.

Archer, S.G. and Adkins, R. (1999). *IMPRINT User's Guide*. US Army Research Laboratory, Human Research and Engineering Directorate.

Arrow, K.J. (1950). A difficulty in the concept of social welfare. *Journal of Political Economy* 58(4), 328–346.

Ashby, W.R. (1956). *An Introduction to Cybernetics*. London: Chapman and Hall.

Bacon, F. (1620). *Novum Organum Scientiarum*. London.

Barr, N. (Ed.) (2012). The relevance of efficiency to different theories of society. In *Economics of the Welfare State* (5th edn). Oxford/New York: Oxford University Press, p. 46.

Bayes, T. (1763). *An Essay Towards Solving a Problem in the Doctrine of Chances* (read to the Royal Society of England).

Berg, S.L. and Sheridan, T.B. (1985). *Effect of Time Span and Task Load on Pilot Mental Workload*. Man-Machine Systems Laboratory Report. Cambridge, MA: MIT Press.

Birkhoff, G. (1948). *Lattice Theory* (Revised). Providence, RI: American Mathematical Society.

Modeling Human–System Interaction: Philosophical and Methodological Considerations, with Examples, First Edition. Thomas B. Sheridan.
© 2017 John Wiley & Sons, Inc. Published 2017 by John Wiley & Sons, Inc.

Bisantz, A.M., Kirlik, A., Gay, P., Phipps, D.A., Walker, N. and Fisk, A.D. (2000). Modeling and analysis of a dynamic judgment task using a lens model approach. *IEEE Transactions on Systems, Man, and Cybernetics* 30(6), 605–616.

Bloom, B.S., Engelhart, M.D., Furst, E.J., Hill, W.H. and Krathwohl, D.R. (1956). *Taxonomy of Educational Objectives: The Classification of Educational Goals. Handbook I: Cognitive Domain.* New York: David McKay Company.

Brunswik, E. (1956). Representative design and probabilistic theory in a functional psychology. *Psychological Review* 62, 193–217.

Buckner, D.N. and McGrath, J.J. (1963). *Vigilance: A Symposium.* New York: McGraw Hill.

Buharali, A. and Sheridan, T.B. (1982). Fuzzy set aids for telling a computer how to decide. In *Proceedings of the 1982 IEEE International Conference on Cybernetics and Society.* CH-1840-8, Seattle, WA.

Byrne, M., Kirk, A. and Fleetwood, M. (2008). An ACT-R approach to closing the loop on computational modeling. In Foyle, D. and Hooey, B. (Eds.), *Human Performance Modeling in Aviation.* New York: CRC Press, pp. 77–103.

Campbell, J. (1949). *The Hero with a Thousand Faces.* Novato, CA: New World Library.

Chapanis, A., Garner, W.R. and Morgan, C.T. (1949). *Applied Experimental Psychology.* New York: John Wiley & Sons.

Chomsky, N. (1957). *Syntactic Structures.* The Hague/Paris: Mouton.

Chu, Y.Y. and Rouse, W.B. (1979). Adaptive allocation of decision-making responsibility between human and computer in multi-task situations. *IEEE Transactions on Systems, Man, and Cybernetics* SMC-9(12), 769–788.

Chu, Y.Y., Steeb, R. and Freedy, A. (1980). *Analysis and Modeling of Information Handling Tasks in Supervisory Control of Advanced Aircraft.* Technical Report PATR-1080-80-6. Woodland Hills, CA: Perceptronics.

Cooksey, R.W. (1996). *Judgment Analysis: Theory, Methods, and Applications.* San Diego, CA: Academic Press.

Cooper, G.E. (1957). Understanding and interpreting pilot opinion. *Aeronautical Engineering Review* 116(3), 47–52.

Craik, K.J.W. (1943). *The Nature of Explanation.* Cambridge: Cambridge University Press.

Craik, K.J.W. (1947). Theory of the human operator in control systems: 1. The operator as an engineering system. *British Journal of Psychology* 38, 56–61.

Darwin, C. (1859). *The Origin of Species.* London: John Murray.

Deutsch, S.E. (1998). Interdisciplinary foundations for multiple-task human performance modeling in OMAR. In *Proceedings of the 20th Annual Meeting of the Cognitive Science Society,* Madison, WI.

Dickinson, E. (1924). *The Complete Poems.* Boston, MA: Little Brown.

Endsley, M.R. (1995a). Toward a theory of situation awareness in dynamic systems. *Human Factors* 37(1), 32–64.

Endsley, M.R. (1995b). Measurement of situation awareness in dynamic systems. *Human Factors* 37(1), 65–84.

Ferrell, W.R. (1965). Remote manipulation with transmission delay. *IEEE Transactions on Human Factors in Electronics* HFE-6(1), 24–32.

Ferrell, W.R. and Sheridan, T.B. (1967). Supervisory control of remote manipulation. *IEEE Spectrum* 4(10), 81–88.

Fitts, P.M (Ed.) (1951). *Human Engineering for an Effective Air Navigation and Traffic Control System*. Washington, DC: National Research Council.

Fitts, P.M. (1954). The information capacity of the human motor system in controlling the amplitude of movement. *Journal of Experimental Psychology* 47, 381–391.

Foyle, D.C. and Hooey, B.L. (Eds.) (2008). *Human Performance Modeling in Aviation*. New York: CRC Press.

Fromm, E. (1941). *Escape from Freedom*. New York: Holt, Rinehart and Winston.

Gardner, H. (2011). *Frames of Mind: The Theory of Multiple Intelligences*. New York: Basis Books.

Gentner, D. and Stevens, A.L. (Eds.) (1983). *Mental Models*. Hillsdale, NJ: Lawrence Erlbaum.

Gibson, J.J. (1979). *Ecological Approach to Visual Perception*. New York: Psychology Press.

Green, D. and Swets, J. (1966). *Signal Detection Theory and Psychophysics*. New York: John Wiley & Sons.

Hancock, P.A. and Chignell, M.H. (1988). Mental workload dynamics in adaptive interface design. *IEEE Transactions on Systems, Man, and Cybernetics* 18(4), 647–658.

Hancock, P.A. and Desmond, P.A. (Eds.) (2001). *Stress, Workload and Fatigue*. Mahwah, NJ: Erlbaum.

Hancock, P.A. and Szalma, J.L. (Eds.) (2008). Performance under stress. In *Stress and Performance*. Burlington, VT: Ashgate.

Hart, S.G. and Sheridan, T.B. (1984). Pilot workload, performance and aircraft control automation. In *Proceedings of NATO/AGARD Conference 371. Human Factors Considerations in High Performance Aircraft*. Neuilly-sur-Seine, France: NATO, pp. 18.1–18.12.

Hick, W.E. (1952). The information capacity of the human motor system in controlling the amplitude of movement. *Journal of Experimental Psychology* 47, 381–391.

Hoffman, R.R., Lee, J.D., Woods, D.D., Shadbolt, N., Miller, J. and Bradshaw, J.M. (2009). The dynamics of trust in cyberdomains. *IEEE Intelligent Systems* 24(6), 5–11.

Howard, R.A. (1966). Information value theory. *IEEE Transactions on Systems Science and Cybernetics* SSC-2, 22–26.

Hume, D. (1739). *Treatise on Human Nature*. London: Printed for John Noon.

Hursch, C.J., Hammond, K.R. and Hursch, J.L. (1964). Some methodological considerations in multiple-cue probability studies. *Psychological Bulletin* 71, 42–60.

Jagacinski, R.J. and Miller, R.A. (1978). Describing the human operator's internal model of a dynamic system. *Human Factors* 20, 425–433.

Johnson, M., Bradshaw, J.M., Feltovich, P.J., Hoffman, R.R., Jonker, C., van Riemsdijk, B. and Sierhuis, M. (2011). Beyond cooperative robotics: the central role of interdependence in coactive design. *IEEE Intelligent Systems* 26(3), 81–88.

Johnson-Laird, P.N. (1983). *Mental Models: Towards a Cognitive Science of Language, Inference, and Consciousness.* Cambridge, MA: Harvard University Press.

Kaber, D.B. and Endsley, M.R. (2003). The effects of level of automation and adaptive automation on human performance, situation awareness and workload in a dynamic control task. *Theoretical Issues in Ergonomic Science*, 1–40.

Kalman, R.E. (1960). A new approach to linear filtering and prediction problems. *Transactions of the ASME—Journal of Basic Engineering* 82D, 33–45.

Karwowski, W. and Mital, A. (1986). *Application of Fuzzy Set Logic to Human Factors.* In Salvendy, G. (Series Editor), Advances in Human Factors/Ergonomics, Vol. 6. New York: Elsevier.

Kirlik, A. (2006). *Methods and Models of Human-Technology Interaction.* New York/Oxford: Oxford University Press.

Kirwan, B. and Ainsworth, L. (Eds.) (1992). *A Guide to Task Analysis.* New York: Taylor and Francis.

Kleinman, D.L., Baron, S. and Levison, W.H. (1971). A control theoretic approach to manned vehicle systems analysis. *IEEE Transactions on Automatic Control* AC-16, 824–832.

Kosco, B. (1993). *Fuzzy Thinking: The New Science of Fuzzy Logic.* New York: Hyperion.

Kozinski, E., Pack, R., Sheridan, T., Vruels, D. and Seminara, J. (1982). *Performance Measurement System for Training Simulators.* Report NP-2719. Atlanta, GA: Electric Power Research Institute.

Kuhn, T. (1962). *The Structure of Scientific Revolutions.* Chicago, IL: University of Chicago Press.

Lee, J. and See, K. (2004). Trust in automation: designing for appropriate reliance. *Human Factors* 46(1), 50–80.

Leveson, N.G. (2012). *Engineering a Safer World.* Cambridge, MA: MIT Press.

Levy, P. (1981). *G. E. Moore and the Cambridge Apostles.* Oxford: Oxford University Press.

McRuer, D. and H. Jex, H. (1967). A review of quasi-linear pilot models. *IEEE Transactions on Human Factors in Electronics* HFE-8-3, 231–249.

Meadows, D.H., Meadows, D.L., Randers, J. and Behrens, W.W. (1972). *Limits to Growth.* New York: Universe Books.

Mendel, M. and Sheridan, T.B. (1989). Filtering information from human experts. *IEEE Transactions on Systems, Man, and Cybernetics* 36(1), 6–16.

Moray, N. (Ed.) (1979). *Mental Workload: Its Theory and Measurement.* New York: Plenum Press.

Moray, N. (1986). Monitoring behavior and supervisory control. In Boff, K., Kaufmann, L. and Beatty, N. (Eds.) *Handbook of Perception and Human Performance*. New York: John Wiley & Sons.

Moray, N. (1997). Models of models of mental models. Chapter 22. In Sheridan, T. and van Lunteren, T. (Eds.) *Perspectives on the Human Controller*. Mahwah, NJ: Lawrence Erlbaum.

Moray, N. and Inagaki, T. (2014). *A Quantitiative Model of Dynamic Visual Attention*. Technical Report in Researchgate.

Moray, N., Sanderson, P., Shiff, B., Jackson, R., Kennedy, S. and Ting, L. (1982). A model and experiment for the allocation of man and computer in supervisory control. In *Proceedings of IEEE International Conference on Cybernetics and Society*. New York: IEEE, pp. 354–358.

Moray, N., Groeger, J. and Stanton, N. (2016). Quantitative modeling in cognitive ergonomics: predicting signals passed at danger. *Ergonomics* 20, 1–15.

Nickerson, R.S. (1992). *Looking Ahead: Human Factors Challenges in a Changing World*. Hillsdale, NJ: Lawrence Erlbaum.

Nickerson, R.S. (1998). Confirmation bias: a ubiquitous phenomenon in many guises. *Review of General Psychology* 2(2), 175–220.

Norman, D. (1983). Some issues on mental models. In Gentner, D. and Stevens, A. (Eds.) *Mental Models*. Hillsdale, NJ: Erlbaum, pp. 7–14.

Ockham's (1495). Ockham's razor is attributed to his discussion in Book II of Ockham's *Commentary on the Sentences of Peter Lombard*, the latter being a Biblical commentary.

O'Donnell, R.D. and Eggemeier, F.T. (1986). Workload assessment methodology. In Boff, K., Kaufman, L. and Thomas, J. (Eds.) *Handbook of Perception and Performance*, Vol. 1. New York: John Wiley & Sons.

Parasuraman, R., Sheridan, T.B. and Wickens, C.D. (2000). A model for types and levels of human interaction with automation. *IEEE Transactions on Systems, Man, and Cybernetics* 30(3), 286–297.

Peirce, C.S. (2001). *Stanford Encyclopedia of Philosophy*. Stanford, CA: Stanford University Press.

Pinedo, M.L. (2012). *Scheduling: Theory, Algorithms and Systems*. New York: Springer.

Pinet, J. (2016). *Facing the Unexpected in Flight*. New York: CRC Press.

Popper (1997). *Stanford Encyclopedia of Philosophy*. Palo Alto, CA: Stanford University Press.

Raiffa, H. and Schlaifer, R. (1961). *Applied Statistical Decision Theory*. Cambridge, MA: Harvard University School of Business Administration.

Rasmussen, J. (1983). Skills, rules and knowledge: signals, signs and symbols, and other distinctions in human performance models. *IEEE Transactions on Systems, Man, and Cybernetics* SMC-133, 257–267.

Rasmussen, J., Pejtersen, A.M. and Goodstein, L.P. (1994). *Cognitive Systems Engineering*. New York: John Wiley & Sons.

Reason, J. (1991). *Human Error*. Cambridge, UK: Cambridge University Press.

Reid, G.B., Shingledecker, C.A. and Eggemeier, F.T. (1981). Application of conjoint measurement to workload scale development. In *Proceedings of Human Factors Society 25th Annual Meeting*. Santa Monica, CA: Human Factors Society, pp. 522–526.

Rouse, W.B. (1977). Human-computer interaction in multi-task situations. *IEEE Transactions on Systems, Man, and Cybernetics* SMC-7(5), 384–392.

Rouse, W.B. (2014). Human interaction with policy flight simulators. *Applied Ergonomics* 45(1), 72–77.

Rouse, W.B. (2015). When the system is an enterprise. *IEEE Systems, Man, and Cybernetics Magazine* 1(4), 16–19.

Rouse, W.B. and Morris, N.M. (1986). On looking into the black box: prospects and limits in the search for mental models. *Psychological Bulletin* 100(3), 349–363.

Rouse, W.B., Pennock, M.J., Oghbaie, M. and Liu, C. (2016). Interactive visualizations for data support. *Report of the National Center for Biotechnology Information*. Available at: www.ncbi.nlm.nih.gov (accessed July 28, 2016).

Ruffell-Smith, H.P.A. (1979). *A Simulator Study of the Interaction of Pilot Workload with Error, Vigilance and Decisions*. NASA TM–78482. Moffett Field, CA: NASA Ames Research Center.

Sanderson, P.M. (1989). Verbalizable knowledge and skilled task performance: association, dissociation, and mental models. *Journal of Experimental Psychology: Learning, Memory, and Cognition* 15(4), 729–747.

Schaefer, K.E., Chen, J.Y.C., Szalma, J.L. and Hancock, P.A. (2016). A meta-analysis of factors influencing the development of trust in automation. *Human Factors* 58(3), 377–400.

Schiele, F. and Green, T. (1990). HCI formalisms and cognitive psychology: the case of task-action grammar. In Harrison, M. and Thimbleby, H. (Eds.) *Formal Methods in Human-Computer Interaction*. Cambridge/New York: Cambridge University Press, pp. 9–62.

Senders, J.W., Elkind, J.I., Grignetti, M.C. and Smallwood, R.P. (1964). *An Investigation of the Visual Sampling Activity of Human Observers (NASA-CR-434)*. Cambridge, MA: Bolt, Beranek and Newman.

Shafer, G. (1976). *A Mathematical Theory of Evidence*. Princeton, NJ: Princeton University Press.

Shannon, C.E. (1949). Communication in the presence of noise. *Proceedings of the IRE* 37, 10–22.

Sheridan, T.B. (1966). Three models of preview control. *IEEE Transactions on Human Factors in Electronics* HFE-6(June), 91–102.

Sheridan, T.B. (1970a). How often the supervisor should sample. *IEEE Transactions on Systems Science and Cybernetics* SSC-6(2), 140–145.

Sheridan, T.B. (1970b). Optimum allocation of personal presence. *IEEE Transactions on Systems Science and Cybernetics* HFE-10, 242–249.

Sheridan, T.B. (1981). Understanding human error and aiding human diagnostic behavior in nuclear power plants. In Rasmussen, J. and Rouse, W.B. (Eds.) *Human Detection and Diagnosis of System Failures.* New York: Plenum Press.

Sheridan, T.B. (1992a). *Telerobotics, Automation and Human Supervisory Control.* Cambridge, MA: MIT Press.

Sheridan, T.B. (1992b). Musings on telepresence and virtual presence. *Presence Teleoperators and Virtual Environments* 1, 120–125.

Sheridan, T.B. (1996). Further musings on the psychophysics of presence. *Presence Teleoperators and Virtual Environments* 5, 241–246.

Sheridan, T.B. (2007). *Vehicle operations simulator with augmented reality.* US Patent Office, Patent Number 7246050.

Sheridan, T.B. (2011). Adaptive automation, level of automation, allocation authority, supervisory control, and adaptive control: distinctions and modes of adaptation. *IEEE Transactions on Systems, Man, and Cybernetics* 41(4), 662–667.

Sheridan, T.B. (2013). Human response is lognormal: plan on waiting if you want reliability. *Ergonomics in Design* 21(1), 4–6.

Sheridan, T.B. (2014). *What is God? Can Religion be Modeled?* Washington, DC: New Academia.

Sheridan, T.B. (2016). Human–robot interaction: status and challenges. *Human Factors* 58(4), 525–532.

Sheridan, T. and Ferrell, W.R. (1974). *Man-Machine Systems.* Cambridge, MA: MIT Press.

Sheridan, T.B. and Johannsen, G. (1976). *Monitoring Behavior and Supervisory Control.* New York: Plenum Press.

Sheridan, T.B. and Simpson, R.W. (1979). *Toward the Definition and Measurement of the Mental Workload of Transport Pilots.* MIT Flight Technology Laboratory Report R79–4, January.

Sheridan, T.B. and Verplank, W.L. (1978). *Human and Computer Control of Undersea Teleoperators.* Man-Machine Systems Laboratory Report. Cambridge, MA: MIT Press.

Skinner, B.F. (1938). *The Behavior of Organisms: An Experimental Analysis.* New York: Appleton Century Crofts.

Sterman, J.D. (2001). System dynamics modeling: tools for learning in a complex world. *California Management Review* 43(4), 8–25.

Stevens, S.S. (1951). *Handbook of Experimental Psychology.* New York: John Wiley & Sons (see chapter 1).

Stone, A.A. and Mackie, C. (Eds.) (2013). *Subjective Well Being in a Policy Relevant Framework.* National Research Council. Washington, DC: National Academies Press.

Taylor, F.W. (1947). *Scientific Management.* New York: Harper and Row.

Thompson, D.A. (1977). The development of a six degree-of-freedom robot evaluation test. In *Proceedings of the 13th Annual Conference on Manual Control.* Cambridge, MA: MIT Press.

Tomizuka, M. (1975). Optimal continuous finite preview problems. *IEEE Transactions on Automatic Control* AC-20(3), 362–365.

Tsach, U., Sheridan, T. and Tzelgov, J. (1982). A new method for failure detection and location in complex systems. In *Proceedings of the IEEE American Control Conference*, Arlington, VA.

Tulga, M.K. and Sheridan, T.B. (1980). Dynamic decisions and workload in multi-task supervisory control. *IEEE Transactions on Systems, Man, and Cybernetics* SMC-10(5), 217–231.

U.S. Supreme Court (1993). *Opinions*, United States Reports. Washington, DC: U.S. Supreme Court.

Verplank (1977). Is there an optimal workload in manual control? PhD Thesis, MIT.

Von Neumann, J. and Morganstern, O. (1944). *Theory of Games and Economic Behavior*. Princeton, NJ: Princeton University Press.

Wickens, C.D. (1984). Processing resources in attention. In Parasuraman, R. and Davies, D.R. (Eds.) *Varieties of Attention*. New York: Academic Press, pp. 63–102.

Wickens, C.D. (2002). Multiple resources and performance prediction. *Theoretical Issues in Ergonomic Science* 3(2), 159–177.

Wickens, C.D., Tsang, P. and Pierce, B. (1985). The dynamics of resource allocation. In Rouse, W.B. (Ed.) *Advances in Man-Machine Systems Research*, Vol. 2. Greenwich, CT: JAI Press.

Wickens, C.D., Goh, J., Helleberg, J., Horrey, W.J. and Talleur, D.A. (2003). Attentional models of multitask pilot performance using advanced display technology. *Human Factors* 45(3), 360–380.

Wickens, C.D., Sebok, A., Gore, B.F. and Hooey, B.L. (2012). Predicting pilot error in NextGen: pilot performance modeling and validation efforts. In *Proceedings of the 4th International Conference on Applied Human Factors and Ergonomics (AHFE)*. Santa Monica, CA: Human Factors and Ergonomics Society.

Wiener, N. (1964). *God and Golem, Incorporated*. Cambridge, MA: MIT Press.

Wierwille, W.W., Casali, J.G., Connor, S.A. and Rahimi, M. (1985). Evaluation of the sensitivity and intrusion of mental workload estimation techniques. In Rouse, W.B. (Ed.) *Advances in Man–Machine Systems Research*, Vol. 2. Greenwich, CT: JAI Press, pp. 51–127.

Woods, D.D., Levison, N. and Hollnagel, E. (2012). *Resilience Engineering: Concepts and Precepts*. Burlington, VT: Ashgate.

Yerkes, R.M. and Dodson, J.D. (1908). The relation of the strength of stimulus to rapidity of habit formation. *Journal of Comparative Neurology and Psychology* 18, 459–482.

Yufik, Y.M. (2013). Towards a theory of understanding and mental modeling. *Recent Advances in Computer Science*, 250–255.

Zadeh, L. (1965). Fuzzy sets. *Information and Control* 8, 338–353.

INDEX

Modeling Human–System Interaction: Philosophical and Methodological Considerations, with Examples,
First Edition. Thomas B. Sheridan.
© 2017 John Wiley & Sons, Inc. Published 2017 by John Wiley & Sons, Inc.